T0296328

A first course in algebraic topology

CZES KOSNIOWSKI

A FIRST COURSE IN
algebraic topology

*Czes Kosniowski is Lecturer in Mathematics
The University of Newcastle upon Tyne*

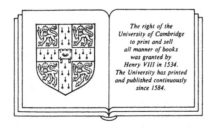

The right of the
University of Cambridge
to print and sell
all manner of books
was granted by
Henry VIII in 1534.
The University has printed
and published continuously
since 1584.

CAMBRIDGE UNIVERSITY PRESS
CAMBRIDGE
LONDON NEW YORK NEW ROCHELLE
MELBOURNE SYDNEY

CAMBRIDGE UNIVERSITY PRESS
Cambridge, New York, Melbourne, Madrid, Cape Town, Singapore, São Paulo, Delhi

Cambridge University Press
The Edinburgh Building, Cambridge CB2 8RU, UK

Published in the United States of America by Cambridge University Press, New York

www.cambridge.org
Information on this title: www.cambridge.org/9780521231954

© Cambridge University Press 1980

First published 1980
Reprinted 1987
Re-issued in this digitally printed version 2008

A catalogue record for this publication is available from the British Library

ISBN 978-0-521-23195-4 hardback
ISBN 978-0-521-29864-3 paperback

CONTENTS

Contents

PREFACE

This book provides a variety of self-contained introductory courses on algebraic topology for the average student. It has been written with a geometric flavour and is *profusely illustrated* (after all, topology is a branch of geometry). Abstraction has been avoided as far as possible and in general a pedestrian approach has been taken in introducing new concepts. The prerequisites have been kept to a minimum and no knowledge of point set or general topology is assumed, making it especially suitable for a first course in topology with the main emphasis on algebraic topology. Using this book, a lecturer will have much freedom in designing an undergraduate or low level postgraduate course.

Throughout the book there are numerous exercises of varying degree to aid and tax the reader. It is, of course, advisable to do as many of these exercises as possible. However, it is not necessary to do any of them, because rarely at any stage is it assumed that the reader has solved the exercises; if a solution to an exercise is needed in the text then it is usually given.

The contents of this book contain topics from topology and algebraic topology selected for their 'teachability'; these are possibly the more elegant parts of the subject. Ample suggestions for further reading are given in the last chapter.

Roughly one-quarter of the book is on general topology and three-quarters on algebraic topology. The general topology part of the book is not presented with its usual pathologies. Sufficient material is covered to enable the reader to quickly get to the 'interesting' part of topology. In the algebraic topology part, the main emphasis is on the fundamental group of a space. Students tend to grasp the concept of the fundamental group readily and it provides a good introduction to what algebraic topology is about. The theory of covering spaces and the Seifert–Van Kampen theorem are covered in detail and both are used to calculate fundamental groups. Other topics include manifolds and surfaces, the Jordan curve theorem (as an appendix to

Chapter 12), the theory of knots and an introductory chapter on singular homology.

As this book is about topology, and not the history of topology, names and dates have not always been included.

This book should not necessarily be read in a linear fashion. The following chart shows the approximate interdependence of the various chapters. For example, to understand Chapter 18 completely you ought to have read Chapters 0-9, 12-16 and 17 beforehand.

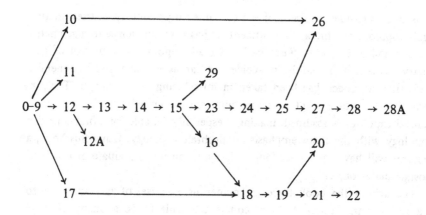

Czes Kosniowski
Newcastle-upon-Tyne
September 1979

0

Sets and groups

In this chapter we give some of the basic definitions and results of set theory and group theory that are used in the book. It is best to refer back to this chapter when the need arises.

For sets X, Y we use the notation $Y \subseteq X$ to mean that Y is a subset of X and $Y \subset X$ to mean that Y is a subset of X and $Y \neq X$. If $Y \subseteq X$ then we denote by X-Y the set of the elements of X which do not belong to Y. The empty set is denoted by $\emptyset$.

The *cartesian* or *direct* product of two sets X and Y is the set of ordered pairs of the form (x, y) where $x \in X$ and $y \in Y$, i.e.

$$X \times Y = \{ (x, y); x \in X, y \in Y \} .$$

The cartesian product of a finite collection $\{ X_i; i=1, 2, ..., n \}$ of sets can be defined analogously:

$$X_1 \times X_2 \times ... \times X_n = \{ (x_1, x_2, ..., x_n); x_i \in X_i, 1 \leq i \leq n \} .$$

A *function* or *map* f: $X \to Y$ between two sets is a correspondence that associates with each element x of X a unique element f(x) of Y. The *identity function* on a set X is the function 1: $X \to X$ such that 1(x) = x for all $x \in X$. The *image* of the function f: $X \to Y$ is defined by

$$Im(f) = f(X) = \{ y \in Y; y=f(x) \text{ for some } x \in X \} .$$

Note that if W, W' are two subsets of X then

$$f(W \cup W') = f(W) \cup f(W'),$$
$$f(W \cap W') \subseteq f(W) \cap f(W').$$

More generally, if we have a collection of subsets of X, say $\{ W_j; j \in J \}$ where J is some *indexing* set, then

$$f(\bigcup_{j \in J} W_j) = \bigcup_{j \in J} f(W_j),$$
$$f(\bigcap_{j \in J} W_j) \subseteq \bigcap_{j \in J} f(W_j).$$

We often abbreviate f: X → Y simply by f if no confusion can arise. A function f: X → Y defines a function from X to f(X) which is also denoted by f. If A is a subset of X then f *restricted* to A is denoted by f|A; it is the function f|A : A → Y defined by (f|A)(a) = f(a) for a ∈ A.

If Z is a subset of Y and f: X → Y is a function then the *inverse image* of Z under f is

$$f^{-1}(Z) = \{\ x \in X; f(x) \in Z\ \}\ .$$

Note that

$$f^{-1}(\underset{j \in J}{\cup}\ Z_j) = \underset{j \in J}{\cup}\ f^{-1}(Z_j)$$

$$f^{-1}(\underset{j \in J}{\cap}\ Z_j) = \underset{j \in J}{\cap}\ f^{-1}(Z_j),$$

$$f^{-1}(Y\text{-}Z_j) = X\text{-}f^{-1}(Z_j),$$

for a collection $\{\ Z_j; j \in J\ \}$ of subsets Z_j of Y.

A function f: X → Y is *one-to-one* or *injective* if whenever $x_1, x_2 \in X$ with $x_1 \neq x_2$ then $f(x_1) \neq f(x_2)$. A function f: X → Y is *onto* or *surjective* if f(X) = Y. A function f: X → Y that is both injective and surjective is said to be *bijective*. In this case there is an *inverse function* f^{-1}: Y → X defined by

$$x = f^{-1}(y) \Leftrightarrow y = f(x).$$

If f: X → Y and g: Y → Z are functions then the *composite function* gf: X → Z is defined by

$$gf(x) = g(f(x)), x \in X.$$

If f: X → Y is a bijective function then ff^{-1}: Y → Y and $f^{-1}f$: X → X are the identity functions. Conversely if gf: X → X and fg: Y → Y are the identity functions then f and g are bijective functions, each being the inverse of the other. The condition that gf: X → X is the identity function implies that f is injective and g is surjective.

A *relation* on a set X is a subset ~ of X × X. We usually write x ~ y if (x,y) ∈ ~. A relation ~ on X is an *equivalence relation* if it satisfies the following three conditions.

(i)   The reflexive condition: x ~ x for all x ∈ X.

(ii)   The symmetric condition: If x ~ y then y ~ x.

(iii)   The transitive condition: If x ~ y and y ~ z then x ~ z.

The *equivalence class* of x is the set

$$[x] = \{\ y \in X; x. \sim y\ \}\ .$$

If ~ is an equivalence relation on X then each element of X belongs to precisely one equivalence class.

A *binary operation* on a set X is a function f: X X X → X. We abbreviate f(x,y) to xy (multiplicative notation) or occasionally x + y (additive notation).

A *group* is a set G together with a binary operation satisfying three conditions:

(1)   There exists an element $1 \in G$, the *identity element* of G, such that $g1 = 1g = g$ for all $g \in G$.

(2)   For each $g \in G$ there is an element $g^{-1} \in G$, the *inverse* of g, such that $gg^{-1} = g^{-1}g = 1$.

(3)   For all $g_1, g_2, g_3 \in G$ *associativity* holds, i.e.

$$(g_1 g_2)g_3 = g_1(g_2 g_3).$$

In the additive group notation the identity element is denoted by 0 and the inverse of g by $-g$. A group whose only element is the identity is the *trivial group* $\{1\}$ or $\{0\}$.

A subset H of a group is a *subgroup* of G if H is a group under the binary operation of G. If H is a subgroup of G and $g \in G$ then the *left coset of* H *by* g is the subset

$$gH = \{gh; h \in H\}.$$

Right cosets are defined analogously. Two left cosets gH, g'H of a subgroup H are either disjoint or identical.

The *direct product* G X H of groups G and H is the set G X H with binary operation defined by (g,h) (g',h') = (gg',hh'). In the additive case we refer to the *direct sum* and denote it by $G \oplus H$.

A *homomorphism* f: G → H from a group G to a group H is a function such that

$$f(gg') = f(g)f(g')$$

for all $g,g' \in G$. If the homomorphism f: G → H is bijective then we say that G and H are *isomorphic groups,* that f is an *isomorphism* and we write $G \cong H$ or f: $G \cong H$. The *kernel* of a homomorphism f: G → H is the set

$$\ker f = \{g \in G; f(g) = 1_H\}$$

where 1_H is the identity of H. The kernel of an isomorphism consists of only the identity element of G.

A subgroup K of a group G is *normal* if $gkg^{-1} \in K$ for all $g \in G$, $k \in K$. The kernel of a homomorphism f: G → H is a normal subgroup of G. A homomorphism f: G → H is injective if and only if ker f = $\{1\}$.

If K is a normal subgroup of G then the left coset gK equals the right coset Kg and the set G/K of all left cosets of K is a group under the operation

$$(gK)(g'K) = (gg')K.$$

We call G/K the *quotient group* of G by K.

The *first isomorphism theorem* states that if f: G → H is a surjective homo-morphism from a group G to a group H with kernel K then H is isomorphic to the quotient group G/K.

If g ∈ G then the *subgroup generated by g* is the subset of G consisting of all integral powers of g

$$\langle g \rangle = \{ g^n ; n \in \mathbb{Z} \}$$

where $g^n = \overbrace{gg \dots g}^{n}$ if $n \geq 0$ and $g^n = \overbrace{g^{-1}g^{-1} \dots g^{-1}}^{-n}$ if $n \leq 0$. In the case of additive notation we have

$$\langle g \rangle = \{ ng; n \in \mathbb{Z} \}$$

where $ng = \overbrace{g+g+ \dots +g}^{n}$ if $n \geq 0$ and $ng = \overbrace{-g+(-g)+ \dots +(-g)}^{-n}$ if $n \leq 0$. If $G = \langle g \rangle$ for some g then we say that G is a *cyclic group* with generator g. In general a *set of generators* for a group G is a subset S of G such that each element of G is a product of powers of elements taken from S. If S is finite then we say that G is *finitely generated*.

A group G is said to be *abelian* or *commutative* if $gg' = g'g$ for all $g,g' \in G$. For example, the set of integers $\mathbb{Z}$ is an abelian group (additive notation); moreover it is a cyclic group generated by +1 or -1.

A *free abelian group of rank n* is a group isomorphic to $\mathbb{Z} \oplus \mathbb{Z} \oplus \dots \oplus \mathbb{Z}$ (n copies).

The *decomposition theorem for finitely generated abelian groups* states: If G is a finitely generated abelian group then G is isomorphic to

$$H_0 \oplus H_1 \oplus \dots \oplus H_m$$

where H_0 is a free abelian group and the H_i, i=1,2,...,m, are cyclic groups of prime power order. The rank of H_0 and the orders of the cyclic subgroups $H_1, H_2,...,H_m$ are uniquely determined.

A commutator in a group G is an element of the form $ghg^{-1}h^{-1}$. The *commutator subgroup* of G is the subset of G consisting of all finite products of commutators of G (it is a subgroup). The commutator subgroup K is a normal subgroup of G and it is in fact the smallest subgroup of G for which G/K is abelian.

We use $\mathbb{R}, \mathbb{C}, \mathbb{Z}, \mathbb{N}, \mathbb{Q}$ to denote the set of real numbers, complex num-bers, integers, natural numbers (or positive integers) and rational numbers respectively. We often refer to $\mathbb{R}$ as the real line and to $\mathbb{C}$ as the complex plane. The set $\mathbb{R}^n$ is the cartesian product of n copies of $\mathbb{R}$. We use the following notation for certain subsets of $\mathbb{R}$ (called *intervals*):

$$(a,b) = \{\, x \in \mathbb{R}\,; a < x < b \,\},$$
$$[a,b] = \{\, x \in \mathbb{R}\,; a \leq x \leq b \,\},$$
$$[a,b) = \{\, x \in \mathbb{R}\,; a \leq x < b \,\},$$
$$(a,b] = \{\, x \in \mathbb{R}\,; a < x \leq b \,\}.$$

The meaning of the subsets $(-\infty,b)$, $(-\infty,b]$, $[a,\infty)$ and (a,∞) should be apparent. Observe that $(-\infty,\infty) = \mathbb{R}$.

Note that (a,b) could refer to a pair of elements, say in $\mathbb{R}^2$ for example, as well as an interval in $\mathbb{R}$. What is meant in a particular instance should be clear from the context.

1

Background: metric spaces

In topology we study sets with some 'structure' associated with them which enable us to make sense of the question Is f: $X \to Y$ continuous or not?, where f: $X \to Y$ is a function between two such sets. In this chapter we shall find out what this 'structure' is by looking at euclidean and metric spaces.

Recall that for a function f: $\mathbb{R} \to \mathbb{R}$ we say that f is *continuous at x* if for all $\epsilon_x > 0$ there exists $\delta_x > 0$ such that $|f(y) - f(x)| < \epsilon_x$ whenever $|y - x| < \delta_x$. The function is then said to be *continuous* if it is continuous at all points $x \in \mathbb{R}$. We can extend this definition of continuity to functions f: $\mathbb{R}^n \to \mathbb{R}^m$ simply by replacing the modulus sign by the euclidean distance. More generally, if we have sets with 'distance functions' then we can define continuity using these distance functions. A 'distance function' – properly called a metric – has to satisfy some (obvious) conditions and these lead to a definition.

1.1   Definition
Let A be a set. A function d: $A \times A \to \mathbb{R}$ satisfying
(i) $d(a,b) = 0$ if and only if $a = b$,
(ii) $d(a,b) + d(a,c) \geq d(b,c)$ for all $a,b,c \in A$
is called a *metric* for A. A set A with a particular metric on it is called a *metric space* and is denoted by (A,d) or simply M.

The second property is known as the *triangle inequality*.

1.2   Exercise
Show that if d is a metric for A then $d(a,b) \geq 0$ and $d(a,b) = d(b,a)$ for all $a,b \in A$.

If we take $A = \mathbb{R}$ and $d(x,y) = |x - y|$ then it is not difficult to see that d is a metric. More generally take $A = \mathbb{R}^n$ and define d by

$$d(x,y) = \left(\sum_{i=1}^{n} (x_i - y_i)^2 \right)^{\frac{1}{2}} = \|x-y\|$$

where $x = (x_1, x_2, ..., x_n)$ and $y = (y_1, y_2, ..., y_n)$. Again, it is not hard to show that d is a metric for $\mathbb{R}^n$. This metric is called the *euclidean* or *usual* metric.

Two other examples of metrics on $A = \mathbb{R}^n$ are given by

$$d(x,y) = \sum_{i=1}^{n} |x_i - y_i|, \quad d(x,y) = \max_{1 \le i \le n} |x_i - y_i|.$$

We leave it as an exercise for the reader to check that these do in fact define a metric.

Finally, if A is any set then we can define a metric on it by the rules $d(x,y) = 0$ if $x = y$ and $d(x,y) = 1$ if $x \ne y$. The resulting metric is called the *discrete metric* on A.

1.3 Exercises

(a) Show that each of the following is a metric for $\mathbb{R}^n$:

$$d(x,y) = \left(\sum_{i=1}^{n} (x_i - y_i)^2 \right)^{\frac{1}{2}} = \|x-y\| ; \quad d(x,y) = \begin{cases} 0 \text{ if } x=y, \\ 1 \text{ if } x \ne y; \end{cases}$$

$$d(x,y) = \sum_{i=1}^{n} |x_i - y_i| ; \quad d(x,y) = \max_{1 \le i \le n} |x_i - y_i|.$$

(b) Show that $d(x,y) = (x - y)^2$ does not define a metric on $\mathbb{R}$.

(c) Show that $d(x,y) = \min_{1 \le i \le n} |x_i - y_i|$ does not define a metric on $\mathbb{R}^n$.

(d) Let d be a metric and let r be a positive real number. Show that d_r defined by $d_r(x,y) = rd(x,y)$ is also a metric.

(e) Let d be a metric. Show that d' defined by

$$d'(x,y) = \frac{d(x,y)}{1 + d(x,y)}$$

is also a metric.

(f) In $\mathbb{R}^2$ define $d(x,y) =$ smallest integer greater or equal to usual distance between x and y. Is d a metric for $\mathbb{R}^2$?

Continuity between metric spaces, as we have indicated, is now easy to define.

1.4   Definition

Let (A,d_A), (B,d_B) be metric spaces. A function f: $A \to B$ is said to be *continuous* at $x \in A$ if and only if for all $\epsilon_x > 0$ there exists $\delta_x > 0$ such that $d_B(f(x),f(y)) < \epsilon_x$ whenever $d_A(x,y) < \delta_x$. The function is said to be *continuous* if it is continuous at all points $x \in A$.

1.5   Exercises

(a) Let A be a metric space with metric d. Let $y \in A$. Show that the function f: $A \to \mathbb{R}$ defined by $f(x) = d(x,y)$ is continuous where $\mathbb{R}$ has the usual metric.

(b) Let M be the metric space $(\mathbb{R},d)$ where d is the usual euclidean metric. Let M_0 be the metric space $(\mathbb{R},d_0)$ where d_0 is the discrete metric, i.e.

$$d_0(x,y) = \begin{cases} 0 \text{ if } x = y, \\ \\ 1 \text{ if } x \neq y. \end{cases}$$

Show that all functions f: $M_0 \to M$ are continuous. Show that there does not exist any injective continuous function from M to M_0.

It is often true that by changing the metric on A or B we do not change the set of continuous functions from A to B. For examples see the following exercises.

1.6   Exercises

(a) Let A,B be metric spaces with metrics d and d_B respectively. Let d_r be the metric on A as given in Exercise 1.3(d) (i.e. $d_r(x,y) = rd(x,y)$). Let f be a function from A to B. Prove that f is continuous with respect to the metric d on A if and only if it is continuous with respect to the metric d_r on A.

(b) As (a) but replace d_r by the metric d' of Exercise 1.3(e).

So distance is not the important criterion for whether or not a function is continuous. It turns out that the concept of an 'open set' is what matters.

1.7   Definition

A subset U of a metric space (A,d) is said to be *open* if for all $x \in U$ there exists an $\epsilon_x > 0$ such that if $y \in A$ and $d(y,x) < \epsilon_x$ then $y \in U$.

In other words U is open if for all $x \in U$ there exists an $\epsilon_x > 0$ such that $B_{\epsilon_x}(x) = \{ y \in A; d(y,x) < \epsilon_x \} \subseteq U$.

An example of an open set in $\mathbb{R}$ is $(0,1) = \{ x \in \mathbb{R}; 0 < x < 1 \}$. In $\mathbb{R}^2$

the following are open sets:

$$\{ (x,y) \in \mathbb{R}^2 ; x^2 + y^2 < 1 \} , \quad \{ (x,y) \in \mathbb{R}^2 ; x^2 + y^2 > 1 \} ,$$
$$\{ (x,y) \in \mathbb{R}^2 ; 0 < x < 1, 0 < y < 1 \} .$$

1.8 Exercises

(a) Show that $B_\epsilon (x)$ is always an open set for all x and all $\epsilon > 0$.

(b) Which of the following subsets of $\mathbb{R}^2$ (with the usual topology) are open?

$$\{ (x,y); x^2 + y^2 < 1 \} \cup \{ (1,0) \} , \quad \{ (x,y); x^2 + y^2 \leq 1 \} ,$$
$$\{ (x,y); |x| < 1 \} , \qquad\qquad\qquad \{ (x,y); x + y < 0 \} ,$$
$$\{ (x,y); x + y \geq 0 \} , \qquad\qquad\qquad \{ (x,y); x + y = 0 \} .$$

(c) Show that if $\mathscr{F}$ is the family of open sets arising from a metric space then

(i) The empty set $\emptyset$ and the whole set belong to $\mathscr{F}$,

(ii) The intersection of two members of $\mathscr{F}$ belongs to $\mathscr{F}$,

(iii) The union of *any* number of members of $\mathscr{F}$ belongs to $\mathscr{F}$.

(d) Give an example of an infinite collection of open sets of $\mathbb{R}$ (with the usual metric) whose intersection is not open.

Using the concept of an open set we have the following crucial result.

1.9 Theorem

A function $f: M_1 \to M_2$ between two metric spaces is continuous if and only if for all open sets U in M_2 the set $f^{-1}(U)$ is open in M_2.

This result says that f is continuous if and only if *inverse* images of open sets are open. It does *not* say that images of open sets are open.

Proof Let d_1 and d_2 denote the metrics on M_1 and M_2 respectively. Suppose that f is continuous and suppose that U is an open subset of M_2. Let $x \in f^{-1}(U)$ so that $f(x) \in U$. Now, there exists $\epsilon > 0$ such that $B_\epsilon (f(x)) \subseteq U$ since U is open. The continuity of f assures that there is a $\delta > 0$ such that

$$d_1(x,y) < \delta \Rightarrow d_2(f(x), f(y)) < \epsilon,$$

or in other words $f(B_\delta(x)) \subseteq B_\epsilon (f(x)) \subseteq U$ which means that $B_\delta (x) \subseteq f^{-1}(U)$. Since this is so for all $x \in f^{-1}(U)$ it follows that $f^{-1}(U)$ is an open subset of M_1.

Conversely let $x \in M_1$; then for all $\epsilon > 0$ the set $B_\epsilon (f(x))$ is an open subset of M_2 so that $f^{-1}(B_\epsilon(f(x)))$ is an open subset of M_1. But this means that since $x \in f^{-1}(B_\epsilon(f(x)))$ there is some $\delta > 0$ with $B_\delta(x) \subseteq f^{-1}(B_\epsilon(f(x)))$, i.e.

$f(B_\delta(x)) \subseteq B_\epsilon(f(x))$. In other words there is a $\delta > 0$ such that $d_2(f(x),f(y)) < \epsilon$ whenever $d_1(x,y) < \delta$, i.e. f is continuous

This theorem tells us, in particular, that if two metrics on a set give rise to the same family of open sets then any function which is continuous using one metric will automatically be continuous using the other. Thus Exercises 1.6 can be rephrased as 'show that the metrics d, d_r and d' give rise to the same family of open sets'.

1.10   Exercise
Which of the metrics $d(x,y) = \Sigma |x_i - y_i|$, $d(x,y) = \max |x_i - y_i|$ on $\mathbb{R}^n$ gives rise to the same family of open sets as that arising from the usual metric on $\mathbb{R}^n$?

From the above we see that in order to study continuity between metric spaces it is the family of open sets in each metric space that is important, and not the metric itself. This leads to the following idea: Given a set X choose a family $\mathscr{F}$ of subsets of X and call these the 'open sets' of X. This gives us an object $(X,\mathscr{F})$ consisting of a set X together with a family $\mathscr{F}$ of subsets of X. Continuity between two such objects $(X,\mathscr{F})$, $(Y,\mathscr{F}')$ could then be defined by saying that f: $X \rightarrow Y$ is continuous if $f^{-1}(U) \in \mathscr{F}$ whenever $U \in \mathscr{F}'$. Naturally if we allowed arbitrary families then we would not get any interesting mathematics. We therefore insist that the family $\mathscr{F}$ of 'open sets' obeys some simple rules: rules that the family $\mathscr{F}$ of open sets arising from a metric space obey (Exercise 1.8(c)). These are

      (i)   (for convenience) the empty set $\emptyset$ and the whole set belong to $\mathscr{F}$,

      (ii) the intersection of two members of $\mathscr{F}$ belongs to $\mathscr{F}$,

      (iii) the union of *any* number of members of $\mathscr{F}$ belongs to $\mathscr{F}$.

The 'structure' associated with a set X, referred to at the beginning of this chapter, is simply a family $\mathscr{F}$ of subsets of X satisfying the above three properties. This is the starting point of topology.

Topological spaces

A topological space is just a set together with certain subsets (which will be called open sets) satisfying three properties.

2.1 **Definition**

Let X be a set and let $\mathcal{U}$ be a collection of subsets of X satisfying

(i) $\emptyset \in \mathcal{U}$, $X \in \mathcal{U}$,

(ii) the intersection of two members of $\mathcal{U}$ is in $\mathcal{U}$,

(iii) the union of any number of members of $\mathcal{U}$ is in $\mathcal{U}$.

Such a collection $\mathcal{U}$ of subsets of X is called a *topology* for X. The set X together with $\mathcal{U}$ is called a *topological space* and is denoted by $(X,\mathcal{U})$ which is often abbreviated to T or just X. The members $U \in \mathcal{U}$ are called the *open sets* of T. Elements of X are called *points* of T.

Note that condition (ii) implies that the intersection of a finite number of members of $\mathcal{U}$ is in $\mathcal{U}$. If $\mathscr{P}(X)$ denotes the set of all subsets of X then a topology for X is just a choice of $\mathcal{U} \subseteq \mathscr{P}(X)$ which satisfies the conditions (i), (ii) and (iii) above. Different choices give different topologies for X.

It is important to have many examples of topological spaces. As a first example we immediately have from the last chapter that any metric space gives rise to a topological space. The resulting space is said to have the *metric topology* or the *usual topology*. The converse is not true – that is, there are topological spaces which do not arise from any metric space – see Exercise 2.2(c). Topological spaces that arise from metric space are said to be *metrizable*. Note that two metric spaces may give rise to the same topological space.

By considering the extremes of the possible families of subsets of a set X satisfying the conditions for a topological space we get our next two examples. The first is where $\mathcal{U} = \{ \emptyset, X \}$; this obviously gives a topology for any set X, called the *concrete* or *indiscrete* topology for X. The other extreme is to let $\mathcal{U}$ be the set $\mathscr{P}(X)$ of all subsets of X; this clearly gives a topology for X, called the *discrete* topology on X.

2.2   Exercises

(a) Show that if X has the discrete topology then it is metrizable. (Hint: Consider the discrete metric.)

(b) Let X be a topological space that is metrizable. Prove that for every pair a,b of distinct points of X there are open sets U_a and U_b containing a and b respectively, such that $U_a \cap U_b = \emptyset$.

(c) Use (b) above to show that if X has at least two points and has the concrete topology then it is not metrizable.

An interesting example of a topology for a set X is that known as the *finite complement topology*. Here $\mathcal{U}$ is $\emptyset$,X and those subsets of X whose complements are finite. Of course if X itself is finite then this is just the discrete topology for X. If X is infinite we need to check that the family $\mathcal{U}$ satisfies the three conditions for a topology. The first is trivially true. For the second suppose that $U_1, U_2 \in \mathcal{U}$ so that $X-U_1$ and $X-U_2$ are finite. Thus $(X-U_1) \cup (X-U_2)$ is also finite, but this is equal to $X-(U_1 \cap U_2)$ and so $U_1 \cap U_2 \in \mathcal{U}$. For the third condition we just use the fact that $X-(\cup U_j) = \cap(X-U_j)$.
$$\underset{j \in J}{}$$

If X consists of two points $\{a,b\}$ then there are four different topologies that we could put on X, namely:

$$\mathcal{U}_1 = \{\emptyset, X\} \; ; \mathcal{U}_2 = \{\emptyset, \{a\}, X\} \; ; \mathcal{U}_3 = \{\emptyset, \{b\}, X\} \; ;$$
$$\mathcal{U}_4 = \{\emptyset, \{a\}, \{b\}, X\} \, .$$

We know that $\mathcal{U}_1$ and $\mathcal{U}_4$ are topologies and leave the checking that $\mathcal{U}_2$ and $\mathcal{U}_3$ are topologies for the reader. Note that $(X, \mathcal{U}_2)$ and $(X, \mathcal{U}_3)$ are not metrizable.

Other examples of topological spaces are given in the exercises that follow.

2.3   Exercises

In each case (a), (b), (c) below show that $\mathcal{U}$ is a topology for X.

(a) $X = \mathbb{R}$, $\mathcal{U} = \{\emptyset\} \cup \{\mathbb{R}\} \cup \{(-\infty, x); x \in \mathbb{R}\}$.

(b) $X = \mathbb{N}$ = the positive integers = the natural numbers, $\mathcal{U} = \{\emptyset\} \cup \{\mathbb{N}\} \cup \{O_n; n \geq 1\}$ where $O_n = \{n, n+1, n+2, ...\}$.

(c) $X = \mathbb{R}$, $U \in \mathcal{U}$ if and only if U is a subset of $\mathbb{R}$ and for each $s \in U$ there is a $t > s$ such that $[s,t) \subseteq U$, where $[s,t) = \{x \in \mathbb{R}; s \leq x < t\}$.

(d) Determine the number of distinct topologies on a set with three elements.

(e) Show that neither of the following families of subsets of $\mathbb{R}$ are topologies.

$$\mathcal{U}_1 = \{\emptyset\} \cup \{R\} \cup \{(-\infty,x]; x \in R\}.$$
$$\mathcal{U}_2 = \{\emptyset\} \cup \{R\} \cup \{(a,b); a,b \in R, a < b\}.$$

For any subset Y of a topological space X we could look at the largest open set contained in Y; this is denoted by $\overset{\circ}{Y}$ and is called the *interior* of Y. In other words

$$\overset{\circ}{Y} = \underset{j \in J}{\cup} U_j'$$

where $\{U_j; j \in J\}$ is the family of all open sets contained in Y. Obviously $x \in \overset{\circ}{Y}$ if and only if there is an open set $U \subseteq Y$ such that $x \in U$.

For example let I^n be the following subset of R^n:

$$I^n = \{x = (x_1, x_2, ..., x_n) \in R^n; 0 \le x_i \le 1, i = 1,2,...,n\}.$$

If R^n has the usual topology (i.e. the metric topology with the 'usual' metric $d(x,y) = (\Sigma_{i=1}^n (x_i - y_i)^2)^{1/2}$) then the interior of I^n is

$$\overset{\circ}{I}^n = \{x; 0 < x_i < 1, i = 1,2,...,n\}.$$

To see this let $x \in \overset{\circ}{I}^n$ and let $\epsilon = \min \{1-x_i, x_i; i = 1,2,...,n\}$. An open ball $B_\epsilon(x)$ (i.e. $\{y \in R^n; d(y,x) < \epsilon\}$) of radius ϵ about x is contained in $\overset{\circ}{I}^n$ and so $\overset{\circ}{I}^n$ is open. On the other hand if, for some i, $x_i = 1$ or 0 then any ball $B_r(x)$ of radius r about x contains points not in I^n no matter how small r is. Hence such points are not in the interior of I^n.

Complements of open sets have a special name.

2.4   Definition

A subset C of a topological space X is said to be *closed* if and only if X - C is open.

The next result follows easily from set theoretic results on the complements of intersections and the complements of unions.

2.5   Theorem

(i) $\emptyset$, X are closed,

(ii) the union of any pair of closed sets is closed,

(iii) the intersection of any number of closed sets is closed.

The concept of a closed set could be used to define topological spaces.

2.6   Exercises

(a) Let X be a set and let $\mathcal{V}$ be a family of subsets of X satisfying

(i) $\emptyset$, $X \in \mathcal{V}$,

(ii) the union of any pair of elements of $\mathcal{V}$ belongs to $\mathcal{V}$,

(iii) the intersection of any number of elements of $\mathcal{V}$ belongs to $\mathcal{V}$.

Show that $\mathcal{U} = \{ X-V; V \in \mathcal{V} \}$ is a topology for X.

(b) Prove that in a discrete topological space each subset is simultaneously open and closed.

(c) Show that if a topological space has only a finite number of points each of which is closed then it has the discrete topology.

(d) Show that in the topological space $(\mathbb{R}, \mathcal{U})$, where $\mathcal{U}$ is as defined in Exercise 2.3(c), each of the sets [s,t) is both an open and a closed subset.

For any subset Y of a topological space X we could consider the smallest closed set containing Y; this is denoted by $\bar{Y}$ and is called the *closure* of Y. In other words

$$\bar{Y} = \bigcap_{j \in J} F_j$$

where $\{ F_j; j \in J \}$ is the family of all closed sets containing Y. The points that are in $\bar{Y}$ but not Y are often called the *limit points* of Y. The next result gives an alternative description of $\bar{Y}$.

2.7   Lemma

$x \in \bar{Y}$ if and only if for every open set U containing x, $U \cap Y \neq \emptyset$.

Proof Let $x \in \bar{Y}$ and suppose that there exists an open set U containing x with $U \cap Y = \emptyset$. Thus X-U is closed and $Y \subseteq X-U$ so $\bar{Y} \subseteq X-U$. But $x \in \bar{Y}$ and $x \in U$ is then a contradiction.

Conversely suppose that $x \notin \bar{Y}$ so that $x \in X-\bar{Y}$. But $X-\bar{Y}$ is open and $(X-\bar{Y}) \cap \bar{Y} = \emptyset$ so that $(X-\bar{Y}) \cap Y = \emptyset$ is a contradiction.

If we consider $\mathbb{R}$ with its usual topology then the closure of the sets (a,b), [a,b), (a,b] and [a,b] is [a,b].

2.8   Exercises

(a) Let X be $\mathbb{R}$ with its usual topology. Find the closure of each of the following subsets of X:

A = { 1,2,3,... } , B = { x; x is rational } , C = { x; x is irrational } .

(b) Let X be $\mathbb{R}$ with the topology of Exercise 2.3(c). Find the closure of each of the following subsets of X:

(a,b), [a,b), (a,b], [a,b] .

Further properties of the closure of a set are given as exercises.

2.9   **Exercises**

Prove each of the following statements.

(a) If Y is a subset of a topological space X with $Y \subseteq F \subseteq X$ and F closed then $\bar{Y} \subseteq F$.

(b) Y is closed if and only if $Y = \bar{Y}$.

(c) $\bar{\bar{Y}} = \bar{Y}$.

(d) $\overline{A \cup B} = \bar{A} \cup \bar{B}$, $\overline{A \cap B} \subseteq \bar{A} \cap \bar{B}$.

(e) $X - \overset{\circ}{Y} = \overline{(X - Y)}$.

(f) $\bar{Y} = Y \cup \partial Y$ where $\partial Y = \bar{Y} \cap \overline{(X - Y)}$ (∂Y is called the *boundary* of Y).

(g) Y is closed if and only if $\partial Y \subseteq Y$.

(h) $\partial Y = \emptyset$ if and only if Y is both open and closed.

(i) $\partial(\{ x \in R ; a < x < b \}) = \partial(\{ x \in R ; a \leq x \leq b \}) = \{ a,b \}$.

(j) Prove that Y is the closure of some open set if and only if Y is the closure of its interior.

A concept that will be useful later on is that of a 'neighbourhood' of a point.

2.10   **Definition**

Let X be a topological space. A subset $N \subseteq X$ with $x \in N$ is called a *neighbourhood* of x if there is an open set U with $x \in U \subseteq N$.

In particular an open set itself is a neighbourhood of each of its points. More generally a set A with $\overset{\circ}{A} \neq \emptyset$ is a neighbourhood of each of the points in the interior of A. Some simple properties of neighbourhoods are given in the next exercise. (It is possible to use the results in the next exercise to define topologies.)

2.11   **Exercise**

Let X be a topological space. Prove each of the following statements.

(i) For each point $x \in X$ there is at least one neighbourhood of x.

(ii) If N is a neighbourhood of x and $N \subseteq M$ then M is also a neighbourhood of x.

(iii) If M and N are neighbourhoods of x then so is $N \cap M$.

(iv) For each $x \in X$ and each neighbourhood N of x there exists a neighbourhood U of x such that $U \subseteq N$ and U is a neighbourhood of each of its points.

_____

Continuous functions

3.1   Definition

A function f: $X \to Y$ between two topological spaces is said to be _continuous_ if for every open set U of Y the inverse image $f^{-1}(U)$ is open in X.

The most trivial examples of continuous functions are the identity function $1_X: X \to X$ and the constant function $X \to Y$ which sends every point of X to some fixed point of Y.

If we take a space X with the discrete topology then any function f: $X \to Y$ from X to any topological space Y is continuous. This is clear since the inverse image of any subset of Y is open in X. On the other hand if we take Y with the concrete topology then any function f: $X \to Y$ from any topological space X to Y is also continuous; this is easy to see. In fact there is a converse to these two examples given in the next set of exercises.

The next example is of a non-continuous function. Let $X = (\mathbb{R}, \mathcal{U})$ where $\mathcal{U} = \{ \emptyset \} \cup \{ \mathbb{R} \} \cup \{ (-\infty, x); x \in \mathbb{R} \}$ and let f: $X \to X$ be given by $f(x) = x^2$. The function f is not continuous because $f^{-1}((-\infty, y^2)) = (-y, y)$ which does not belong to $\mathcal{U}$. Exactly which functions from X to X are continuous is left in the form of an exercise (Exercise 3.2(d)).

3.2   Exercises

(a) Let X be an arbitrary set and let $\mathcal{U}, \mathcal{U}'$ be topologies on X. Prove that the identity mapping $(X, \mathcal{U}) \to (X, \mathcal{U}')$ is continuous if and only if $\mathcal{U}' \subseteq \mathcal{U}$.

(b) X is a topological space with the property that, for every topological space Y, every function f: $X \to Y$ is continuous. Prove that X has the discrete topology. (Hint: Let Y be the space X but with the discrete topology.)

(c) Y is a topological space with the property that, for every topological space X, every function f: $X \to Y$ is continuous. Prove that Y has the concrete topology. (Hint: Let X be the space Y but with the concrete topology.)

(d) Let X be the real numbers with the topology $\{\emptyset\} \cup \{\mathbb{R}\} \cup \{(-\infty, x); x \in \mathbb{R}\}$. Prove that a function f: $X \rightarrow X$ is continuous if and only if it is non-decreasing (i.e. if $x > x'$ then $f(x) \geq f(x')$) and continuous on the right in the classical sense (i.e. for all $x \in X$ and all $\epsilon > 0$ there exists $\delta > 0$ such that if $x \leq x' < x + \delta$ then $|f(x) - f(x')| < \epsilon$).

There is a characterization of continuous maps in terms of closed sets.

3.3   Theorem
The function f: $X \rightarrow Y$ between topological spaces X,Y is continuous if and only if $f^{-1}(C)$ is closed for all closed subsets C of Y.

Proof Suppose that f is continuous. If C is closed then $Y-C$ is open which means that $f^{-1}(Y-C)$ is open. But $f^{-1}(Y-C) = X - f^{-1}(C)$ and hence $f^{-1}(C)$ is closed. Conversely suppose that U is open in Y so that $Y-U$ is closed and hence $f^{-1}(Y-U) = X - f^{-1}(U)$ is closed, which means that $f^{-1}(U)$ is open and f is continuous.

A function that sends open sets to open sets is said to be *open*. Open mappings are not necessarily continuous. As an example let Y consist of two points $\{a,b\}$ with the discrete topology and let X be the real numbers with the usual topology. The function f: $X \rightarrow Y$ given by

$$f(x) = \begin{cases} a \text{ if } x \geq 0, \\ b \text{ if } x < 0, \end{cases}$$

is an open mapping but is not continuous because $f^{-1}(\{a\})$ is not open in X. Any mapping from a topological space to a discrete topological space is necessarily open.

We say that a map f: $X \rightarrow Y$ is *closed* if the image under f of any closed set is closed. Closed mappings are not necessarily continuous; in fact the example of the open non-continuous function given earlier is also closed. In general a continuous function may be (i) neither open nor closed, (ii) open but not closed, (iii) closed but not open or (iv) both open and closed. As examples we have the following (i) X is a set A with the discrete topology, Y is the set A with the concrete topology and f is the identity function. For (ii) consider $X = \{a,b\}$ with the discrete topology and $Y = \{a,b\}$ with the topology $\{\emptyset, \{a\}, \{a,b\}\}$; then the constant function to $a \in Y$ is open and continuous but not closed. For (iii) take $X = \{a,b\}$ with the discrete topology and $Y = \mathbb{R}$ with the usual topology; then the function f: $X \rightarrow Y$

given by f(a) = 0, f(b) = 1, is continuous and closed but not open. Finally
for (iv) we can take X = Y to be any topological space and f to be the
identity. Of course if we put some further restrictions on f then all four
cases may not arise.

3.4   Exercise

Let f be a continuous function f: X → Y between the topological
spaces X and Y. If f is (a) injective (b) surjective (c) bijective, which of the
four cases (i) f is neither open nor closed, (ii) f is open but not closed, (iii) f
is closed but not open and (iv) f is both open and closed can actually arise?

The next result tells us that the composite of two continuous functions is
continuous. It is remarkably easy to prove.

3.5   Theorem

Let X,Y and Z be topological spaces. If f: X → Y and g: Y → Z are
continuous functions then the composite h = gf: X → Z is also continuous.

Proof If U is open in Z then $g^{-1}(U)$ is open in Y and so $f^{-1}(g^{-1}(U))$ is open
in X. But $(gf)^{-1}(U) = f^{-1}(g^{-1}(U))$.

The next definition tells us when two topological spaces are considered
equivalent; we use the word homeomorphism.

3.6   Definition

Let X and Y be topological spaces. We say that X and Y are *homeomorphic* if there exist inverse continuous functions f: X → Y, g: Y → X (i.e.
$fg = 1_Y$, $gf = 1_X$ and f,g are continuous). We write $X \cong Y$ and say that f and g
are *homeomorphisms* between X and Y.

An equivalent definition would be to require a function f: X → Y which is
(i) bijective, (ii) continuous and (iii) its inverse f^{-1} is also continuous. Thus
a homeomorphism between X and Y is a bijection between the points and
the open sets of X and Y.

Some examples of homeomorphisms can be readily obtained from Chapter 1. For example if X is the topological space arising from a metric space
M with metric d, and if Y is that arising from the metric space M with
metric d' given by $d'(x,y) = d(x,y)/(1 + d(x,y))$, then X and Y are homeomorphic. Another example is to let X be $\mathbb{R}^n$ with the usual metric topology
and to let Y be $\mathbb{R}^n$ with the metric topology obtained from the metric
$d(x,y) = \max |x_i - y_i|$. Again X and Y are homeomorphic. On the other hand

if $X = \mathbb{R}^n$ with the usual topology and $Y = \mathbb{R}^n$ with the discrete topology then X and Y are not homeomorphic.

3.7 Exercises

(a) Give an example of spaces X,Y and a continuous bijection $f: X \to Y$ such that f^{-1} is not continuous.

(b) Let X and Y be topological spaces. Prove that X and Y are homeomorphic if and only if there exists a function $f: X \to Y$ such that (i) f is bijective and (ii) a subset U of X is open if and only if f(U) is open.

(c) Metrics d_1 and d_2 on a set Y are such that, for some positive m and M,

$$md_1(y,y') \leq d_2(y,y') \leq Md_1(y,y')$$

for all $y,y' \in Y$. Show that the two topological spaces arising from these metrics are homeomorphic. (Hint: Consider the identity mapping on Y.)

(d) Let X be a topological space and let G(X) denote the set of homeomorphisms $f: X \to X$. Prove that G(X) is a group. For $x \in X$, let $G_x(X) = \{ f \in G(X); f(x) = x \}$. Prove that $G_x(X)$ is a subgroup of G(X).

Homeomorphism is an equivalence relation, and topology is the study of the equivalence classes. The next three chapters describe ways of producing new topological spaces from old ones.

Induced topology

Let S be a subset of the topological space X. We can give S a topology from that of X.

4.1   Definition
The topology on S *induced* by the topology of X is the family of sets of the form $U \cap S$ where U is an open set in X.

In other words if $\mathcal{U}$ is the family of open sets in X then $\mathcal{U}_S = \{ U \cap S;$ $U \in \mathcal{U} \}$ is the family of open sets in S. To prove that $\mathcal{U}_S$ gives a topology for S we have to check the three conditions for a topology. Since $\emptyset = \emptyset \cap S$ and $S = X \cap S$ we immediately have the first condition. For the second let $U_1 \cap S$ and $U_2 \cap S$ be two elements of $\mathcal{U}_S$; then since $(U_1 \cap S) \cap (U_2 \cap S) = (U_1 \cap U_2) \cap S$ it belongs to $\mathcal{U}_S$. Finally, if $\{ U_j \cap S; j \in J \}$ is an arbitrary set of elements taken from $\mathcal{U}_S$ then $\underset{j \in J}{\cup} (U_j \cap S) = (\underset{j \in J}{\cup} U_j) \cap S$ is in $\mathcal{U}_S$.

The induced topology is sometimes referred to as the *relative* topology. If the subset S of X has the induced topology then we say that S if a *subspace* of X.

For example, if we take the subset [a,b] of $\mathbb{R}$ (with the usual topology) and give it the induced topology then the sets

$$[a,c), \quad a < c < b,$$
$$(d,b], \quad a < d < b,$$
$$(d,c), \quad a \le d < c \le b$$

are open subsets of [a,b]. Note that U open in [a,b] does not imply that U is open in $\mathbb{R}$.

As another example we can give the unit circle S^1 in $\mathbb{R}^2$, the topology induced by the usual topology on $\mathbb{R}^2$. The open sets of S^1 are then unions of 'open arcs' (i.e. arcs with endpoints excluded). More generally we give the standard n-sphere S^n where

$$S^n = \{\, x \in \mathbb{R}^{n+1};\; \sum_{i=1}^{n+1} x_i^2 = 1 \,\},$$

the topology induced by the usual topology on $\mathbb{R}^{n+1}$.

In $\mathbb{R}^{n+1}$ we may consider the subset S given by $x_{n+1} = 0$. If we give S the induced topology (using the usual topology on $\mathbb{R}^{n+1}$) then S is homeomorphic to $\mathbb{R}^n$. The proof is left as an exercise; alternatively see Chapter 6.

It is interesting to look at subspaces of $\mathbb{R}^3$ and try to find which are homeomorphic to each other. For example the intervals [a,b] and [c,d] in $\mathbb{R} \subset \mathbb{R}^3$ are homeomorphic. A homeomorphism f is given by

$$f(x) = c + (d-c)(x-a)/(b-a).$$

It is not difficult to construct an inverse f^{-1} and show that f and f^{-1} are continuous (see also Exercise 4.5(g)). Intuitively we just stretch or shrink the intervals into each other.

Figure 4.1

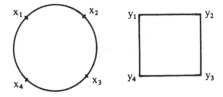

For another example look at a circle and a square (by a square we mean the 'edge' of a square region); see Figure 4.1. The map that sends the intervals in S^1 from x_i to x_{i+1} onto the intervals in the square from y_i to y_{i+1} defines a homeomorphism from the circle to the square. If $\{\, (x,y); x^2 + y^2 = 1 \,\}$ is the circle and $\{\, (x,y); x = \pm 1, -1 \le y \le 1 \text{ or } -1 \le x \le 1, y = \pm 1 \,\}$ is the square then explicit homeomorphisms are given by

circle	→	square		square	→	circle
(x,y)	→	(x/m,y/m)		(x,y)	→	(x/r,y/r)

where m = max (|x|, |y|) and r = $\sqrt{(x^2 + y^2)}$. Intuitively we just twist or bend the circle to form a square. In general if we have two subspaces of $\mathbb{R}^3$ (or $\mathbb{R}^2$) then intuitively they are homeomorphic if we can twist, bend and stretch or shrink one into the other without joining points together and without making any cuts. So, for example, a doughnut (the type with a hole in it) is homeomorphic to a teacup (with a handle); see Figure 4.2.

Another example of homeomorphic spaces is given in Figure 4.3(*a*) and (*d*), with intermediate homeomorphic spaces illustrated in Figure 4.3(*b*) and (*c*).

If h: $X \cong Y$ is a homeomorphism then for every point $x \in X$ the spaces $X - \{ x \}$ and $Y - \{ h(x) \}$ are homeomorphic. This sometimes gives us a way of showing that certain spaces are not homeomorphic. For example, at

Figure 4.2

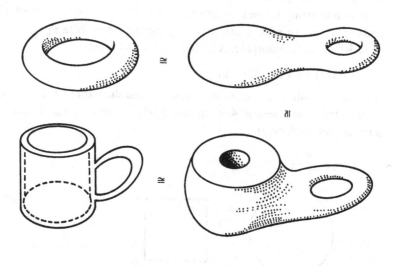

Figure 4.3

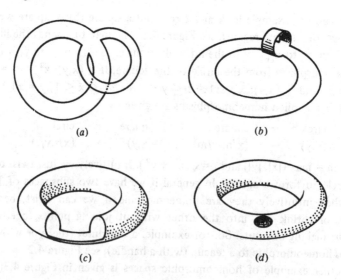

(*a*)   (*b*)

(*c*)   (*d*)

least intuitively at this stage, the subspaces [0,1] and (0,1) of $\mathbb{R}$ are not homeomorphic because if we remove the point 0 from [0,1] then we get (0,1] which (intuitively) is in one piece, whereas if we remove any point from (0,1) then we get (intuitively) two pieces; more precisely it is the disjoint union of two non-empty open subsets. Now (intuitively) one piece cannot be homeomorphic to two pieces (this would involve cutting, which is not continuous) and so [0,1] cannot be homeomorphic to (0,1). (The notion of 'one piece' and 'two pieces' will be made rigorous in Chapter 9.) The above idea can be extended to removing two or more points. The reader can explore these ideas a bit more by doing the exercise that follow.

4.2   Intuitive exercise on homeomorphisms

Sort the subspaces of $\mathbb{R}^3$ (and $\mathbb{R}^2$) in Figure 4.4 into sets of homeomorphic ones.

Figure 4.4

If we look at a circle and a knotted circle in $\mathbb{R}^3$ (see Figure 4.5) then we could easily construct a homeomorphism between the circle and the knotted

Figure 4.5

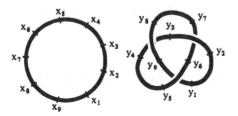

circle. The idea is to divide each into, say nine parts, and map the interval in the circle from x_i to x_{i+1} onto the interval in the knotted circle from y_i to y_{i+1}. If a knotted circle is made out of thin string then, as the reader can easily discover, it is not possible to make a circle from the knotted circle by twisting and bending without cutting or glueing. However, if we make a temporary cut in the knotted circle, unknot it and glue it back we can get a circle. This suggests that we modify our intuitive notion of homeomorphisms between subspaces of $\mathbb{R}^3$ by allowing temporary cuts. The idea is that we can make a temporary cut, carry out some homeomorphisms (by twisting, bending etc.) and then glue back the cut we made; the initial and final spaces are then homeomorphic. This idea can be made rigorous by using the notion of quotient spaces as in Chapter 5; see in particular Theorem 5.5.

4.3   Exercise

Show that the two subspaces of $\mathbb{R}^3$ shown in Figure 4.6 are homeomorphic. The first subspace is obtained by sewing together three twisted strips of paper to two circular discs of paper. The second is obtained by sewing together two long strips of paper. (Hint: Cut the first at two places, namely at two of the twisted strips, then unfold and finally glue back.)

Figure 4.6

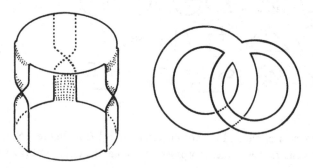

We have already mentioned that if S is a subspace of X then open sets of S are not necessarily open in X. If, however, S is open in X then open subsets of S are open in X.

4.4   Lemma

(i) If S is open in X then the open sets of S in the induced topology are open in X.

(ii) If S is closed in X then the closed sets of S in the induced topology are closed in X.

Proof Since the proofs of (i) and (ii) are more or less identical we shall only
give the proof of (i). Suppose S is open in X and let U be an open subset of S.
By definition $U = V \cap S$ where V is an open subset of X. But since S is open
in X we also have that $U = V \cap S$ is open in X.

4.5   Exercises

(a) Show that if Y is a subspace of X, and Z is a subspace of Y, then Z
is a subspace of X.

(b) Prove that a subspace of a metrizable space is metrizable.

(c) Suppose that S is a subspace of X. Show that the inclusion map
$S \to X$ is continuous. Furthermore, show that S has the weakest
topology (i.e. the least number of open sets) such that the inclusion
$S \to X$ is continuous.

(d) X is a topological space, S is a subset and i: $S \to X$ denotes the
inclusion map. The set S is given a topology such that for every
space Y and map f: $Y \to S$

f: $Y \to S$ is continuous $\Leftrightarrow$ if: $Y \to X$ is continuous.

Prove that the topology on S is the topology induced by the topology on X.

(e) Let Y be a subspace of X and let A be a subset of Y. Denote by
$Cl_X(A)$ the closure of A in X and by $Cl_Y(A)$ the closure of A in Y.
Prove that $Cl_Y(A) \subseteq Cl_X(A)$. Show that in general $Cl_Y(A) \neq Cl_X(A)$.

(f) Show that the subset (a,b) of $\mathbb{R}$ with the induced topology is
homeomorphic to $\mathbb{R}$. (Hint: Use functions like $x \to \tan(\pi(cx + d))$
for suitable c and d.)

(g) Let X,Y be topological spaces and let S be a subspace of X. Prove
that if f: $X \to Y$ is a continuous map then so is f|S: $S \to f(S)$.

(h) Show that the subspaces $(1,\infty)$, $(0,1)$ of $\mathbb{R}$ with the usual topology
are homeomorphic. (Hint: $x \to 1/x$.)

(i) Prove that $S^n - \{ (0,0,...,0,1) \}$ is homeomorphic to $\mathbb{R}^n$ with the
usual topology. (Hint: Define φ: $S^n - \{ (0,0,...,0,1) \} \to \mathbb{R}^n$ by

$$\varphi(x_1,x_2,...,x_{n+1}) = \left(\frac{x_1}{1-x_{n+1}} , \frac{x_2}{1-x_{n+1}} ,..., \frac{x_n}{1-x_{n+1}} \right)$$

and $\psi : \mathbb{R}^n \to S^n - \{ (0,0,...,0,1) \}$ by

$$\psi(x_1,x_2,...,x_n) = \frac{1}{1+\|x\|^2} (2x_1,2x_2,...,2x_n,\|x\|^2 - 1).)$$

(j) Let $\mathbb{R}^{n+1} - \{0\}$ and S^n have the subspace topology of $\mathbb{R}^{n+1}$ with the usual topology. Prove that $f: \mathbb{R}^{n+1} - \{0\} \to S^n$ defined by $f(x) = x/\|x\|$ is a continuous function.

5

Quotient topology (and groups acting on spaces)

In the last chapter we essentially considered a set S, a topological space X and an injective mapping from S to X. This gave us a topology on S: the induced topology. In this chapter we shall consider a topological space X, a set Y and a surjective mapping from X to Y. This will give us a topology on Y: the so-called 'quotient' topology.

5.1   Definition

Suppose that f: X → Y is a surjective mapping from a topological space X onto a set Y. The *quotient topology* on Y with respect to f is the family

$$\mathcal{U}_f = \{ U; f^{-1}(U) \text{ is open in X} \}.$$

It is easy to check that $\mathcal{U}_f$ satisfies the conditions for a topology: obviously $\emptyset \in \mathcal{U}_f$ and $Y \in \mathcal{U}_f$, and the other conditions follow easily from the facts that $f^{-1}(U_1 \cap U_2) = f^{-1}(U_1) \cap f^{-1}(U_2)$ and $f^{-1}(\cup_{j\in J} U_j) = \cup_{j\in J} f^{-1}(U_j)$.

Note that after we give Y the quotient topology then the function f: X → Y is continuous.

A nice example is to take the set $\mathbb{R}P^n = \{ \{ x,-x \} ; x \in S^n \}$ of certain unordered pairs of points in S^n. There is an obvious surjective mapping $\pi: S^n \to \mathbb{R}P^n$ given by $x \to \{ x,-x \}$. The set $\mathbb{R}P^n$ with the quotient topology with respect to the mapping π is called the *real projective n-space*.

As a second example first consider the space

$$C = \{ (x,y,z) \in \mathbb{R}^3 ; x^2 + y^2 = 1, |z| \leq 1 \}$$

with the induced topology. (C is a cylinder.) Let M be the set of unordered pairs of points in C of the form $\{ p,-p \}$, i.e.

$$M = \{ \{ p,-p \} ; p \in C \}.$$

Since we have a natural surjective map from C to M we can give M the quotient topology; the result is called a *Möbius strip* or *band* (sometimes Möbius is spelt Moebius).

Consider the function f: $M \to \mathbb{R}^3$ given by

$$\{p,-p\} \to ((x^2-y^2)(2+xz), 2xy(2+xz), yz)$$

where $p = (x,y,z) \in C \subseteq \mathbb{R}^3$. It is not difficult to check that f is injective. The image f(M) of M under f is pictured in Figure 5.1. In fact M is homeomorphic to f(M) $\subseteq \mathbb{R}^3$ with the induced topology: that f is continuous follows from the fact that F: $\mathbb{R}^3 \to \mathbb{R}^3$ defined by

$$F(x,y,z) = ((x^2-y^2)(2+xz), 2xy(2+xz), yz)$$

is continuous and from the universal mapping property of quotients (Theorem 5.2 below). The fact that f^{-1} is continuous is left for the reader to prove; in fact it follows quite easily from results to be proved in Chapter 8.

We now state and prove the *universal mapping property of quotients*.

5.2   Theorem
Let f: $X \to Y$ be a mapping and suppose that Y has the quotient topology with respect to X. Then a mapping g: $Y \to Z$ from Y to a topological space Z is continuous if and only if gf is continuous.

Proof The function f: $X \to Y$ is continuous and so if g is continuous then so is the composite gf. Conversely suppose that gf is continuous. If V is open in Z then $(gf)^{-1}(V)$ is open in X or in other words $f^{-1}(g^{-1}(V))$ is open in X. By definition of the quotient topology on Y it follows that $g^{-1}(V)$ is open in Y and so g is continuous.

5.3   Exercises
(a) Suppose that Y is given the quotient topology with respect to the mapping f: $X \to Y$. Prove that Y has the strongest topology such that f is continuous.

(b) Suppose that Y has the quotient topology with respect to the mapping f: $X \to Y$. Show that a subset A of Y is closed if and only if $f^{-1}(A)$ is closed in X.

Figure 5.1

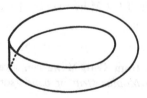

(c) Let f: $\mathbb{R} \to S^1$ ($S^1 \subseteq \mathbb{R}^2$) be defined by

$$f(t) = (\cos(2\pi t), \sin(2\pi t)) \in \mathbb{R}^2.$$

Prove that the quotient topology $\mathscr{U}_f$ on S^1 determined by f is the same as the topology $\mathscr{U}$ induced from $\mathbb{R}^2$ (i.e. show that $(S^1, \mathscr{U}_f)$ $\cong (S^1, \mathscr{U})$).

(d) Let X,Y,Z be topological spaces and let f: $X \to Y$, g: $Y \to Z$ be surjections. Prove that if the topologies of Y and Z are the quotient topologies determined by f and g respectively then the topology of Z is the quotient topology determined by gf: $X \to Z$.

(e) Prove that $\mathbb{R}P^1$ and S^1 are homeomorphic.

(f) Show that the function f: $\mathbb{R}P^2 \to \mathbb{R}^4$ given by

$$\{x,-x\} \to (x_1^2-x_2^2, x_1x_2, x_1x_3, x_2x_3)$$

is continuous and injective.

(g) Let X be a topological space and let f: $X \to Y$ be a surjective map. Let $\mathscr{U}_f$ denote the quotient topology on Y. Suppose that $\mathscr{U}$ is a topology on Y so that f: $X \to Y$ is continuous with respect to this topology. Prove that if f is a closed or an open mapping then $(Y,\mathscr{U})$ is homeomorphic to $(Y,\mathscr{U}_f)$. Furthermore, give examples to show that if f is neither open nor closed then $(Y,\mathscr{U}) \ncong (Y, \mathscr{U}_f)$.

(h) Suppose that f: $X \to Y$ is a surjective map from a topological space X to a set Y. Let Y have the quotient topology determined by f and let A be a subspace of X. Let $\mathscr{U}_1$ denote the topology on B = f(A) $\subseteq$ Y induced by Y and let $\mathscr{U}_2$ denote the quotient topology determined by the map f|A: $A \to B$. Show that $\mathscr{U}_1 \subseteq \mathscr{U}_2$. Give an example to show that in general $\mathscr{U}_1 \neq \mathscr{U}_2$. (Hint: Consider f: $\mathbb{R} \to S^1$ given by $f(t) = \exp(2\pi it)$.) Also, show that if either A is a closed subset of X and f is a closed map or A is an open subset of X and f is an open map then $\mathscr{U}_1 = \mathscr{U}_2$.

Surjective mappings are obtained if we consider the equivalence classes of some equivalence relation. Thus, if X is a topological space and $\sim$ is an equivalence relation on X then we let X/$\sim$ denote the set of equivalence classes and define f: $X \to X/\sim$ by f(x) = [x] the equivalence class containing x. X/$\sim$ with the quotient topology is often said to be obtained from X by *topological identification*. For example if $\sim$ is the equivalence relation on S^n given by $x \sim y$ if and only if $x = \pm y$ then $S^n/\sim$ is of course $\mathbb{R}P^n$. Similarly the same relation on the cylinder C gives C/$\sim$ which is the Möbius strip.

If we take the unit square X = $\{(x,y); 0 \leq x, y \leq 1\}$ in $\mathbb{R}^2$ with the induced topology and define an equivalence relation $\sim$ on X by

$$(x,y) \sim (x',y') \Leftrightarrow (x,y) = (x',y') \text{ or } \{x,x'\} = \{0,1\} \text{ and } y = y'$$

then X/~ with the quotient topology is in fact homeomorphic to the cylinder. A tedious proof could be given now but it will follow much more readily in Chapter 8; intuitively it is clear. We picture X with its equivalence relation as in Figure 5.2(*a*), the arrows indicating which (and in which way) points are to be identified.

Figure 5.2

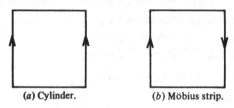

    (*a*) Cylinder.   (*b*) Möbius strip.

We could construct a Möbius strip by a similar process; the relevant picture is given in Figure 5.2(*b*) and the relation on the square X is

$$(x,y) \sim (x',y') \Leftrightarrow (x,y) = (x',y') \text{ or } \{x,x'\} = \{0,1\} \text{ and } y = 1-y'.$$

Two other examples obtained from topological identifications of a unit square are given in Figure 5.3.

Figure 5.3

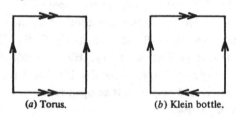

    (*a*) Torus.   (*b*) Klein bottle.

The non-trivial relations on X given in Figure 5.3(*a*) are

$$(0,y) \sim (1,y), \quad (x,0) \sim (x,1),$$

while the non-trivial relations on X given in Figure 5.3(*b*) are

$$(0,y) \sim (1, y), \quad (x,0) \sim (1-x,1).$$

It will be apparent later on (but the reader may like to prove this now) that the torus (the space of Figure 5.3(*a*)) is homeomorphic to the subspace of $\mathbb{R}^3$ given by

$$\{ (..,y,z) \in \mathbb{R}^3 ; (\sqrt{(x^2 + y^2)} - 2)^2 + z^2 = 1 \} .$$

A homeomorphism is given by

$$(x,y) \to ((2+\cos(2\pi x))\cos(2\pi y), (2+\cos(2\pi x))\sin(2\pi y), \sin(2\pi x))$$

This leads to the traditional picture of a torus as the surface of a doughnut (the type with a 'hole' in it). See figure 5.4.

Intuitively, if we start with some flexible material in the shape of Figure 5.3(*a*) and make the appropriate identifications then we are led to this picture again. See Figure 5.5.

The similar process for a Klein bottle is difficult because we need to perform the identification in $\mathbb{R}^4$. The first identification (Figure 5.6(*b*)) is easy. For the second (Figure 5.6(*c*)) we need four dimensions. Pictorially we represent this as in Figure 5.6(*d*). The circle of intersection that appears is

Figure 5.4

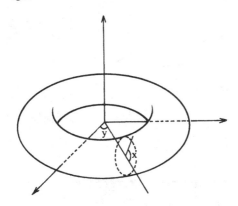

Figure 5.5

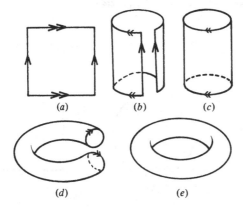

(*a*)   (*b*)   (*c*)

(*d*)   (*e*)

Figure 5.6

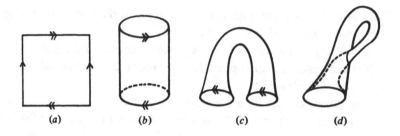

(a)   (b)   (c)   (d)

Figure 5.7

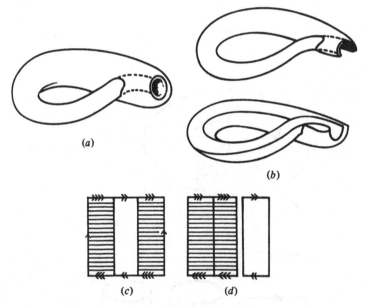

(a)

(b)

(c)   (d)

Figure 5.8

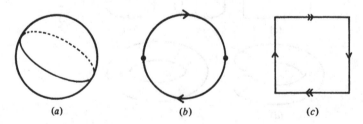

(a)   (b)   (c)

not really present, it appears because we live in a three-dimensional world.

By slicing Figure 5.6(*d*) with a plane we see (see Figure 5.7(*a*),(*b*)) that a Klein bottle is really just two Möbius strips joined along their common boundary. We can also visualize this as in Figure 5.7(*c*),(*d*).

For further intuitive notions recall that the real projective plane $\mathbb{R}P^2$ is defined as $S^2/\sim$ where

$$x \sim x' \Leftrightarrow x = \pm x'.$$

In this case the northern hemisphere is identified with the southern hemisphere and so we may restrict our attention to the northern hemisphere which is homeomorphic to the disc $D^2 = \{ (x,y) \in \mathbb{R}^2 ; x^2 + y^2 \leq 1 \}$ via

$$(x, y, z) \to (x, y)$$

for $(x, y, z) \in S^2$ with $z \geq 0$. Thus we may rewrite $\mathbb{R}P^2$ as $D^2/\sim$ where

$$x \sim x' \Leftrightarrow x = x' \quad \text{or } x, x' \in S^1 \subseteq D^2 \text{ and } x = -x'.$$

Pictorially this gives Figure 5.8(*b*), or equivalently Figure 5.8(*c*). Of course we have not presented a rigorous proof.

If we remove a small region (homeomorphic to $\mathring{D}^2$) in $\mathbb{R}P^2$ then we are left with a Möbius strip; see Figure 5.9. Thus the real projective plane can be thought of as a Möbius strip sewn onto a disc.

Figure 5.9

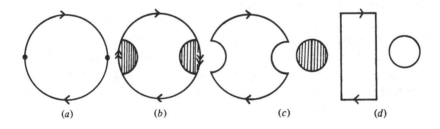

(*a*)　　　(*b*)　　　(*c*)　　　(*d*)

Figure 5.10

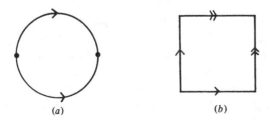

(*a*)　　　(*b*)

The sphere may be represented as a quotient space as indicated in Figure 5.10(*a*) or (*b*). Intuitively we imagine the spaces as purses with a zip. After zipping up we get a sphere.

In the above examples we have argued intuitively. Rigorous proofs could be given but these are best left until we have a bit more theory. After reading Chapter 8 the reader should return to this chapter and give details of proofs concerning the intuitive results just mentioned.

A lower-dimensional analogue of the disc and sphere example is the interval and circle: if we identify the ends of a unit interval we get a circle; intuitively this is clear. The reader should try to write down a proper proof.

5.4   Exercises

(a) Show that if $I = [0,1] \subseteq \mathbb{R}$ and $\sim$ is the equivalence relation $x \sim x'$ if and only if $\{ x,x' \} = \{ 0,1 \}$ or $x = x'$ then $I/\!\sim$ is homeomorphic to S^1.

(b) The Möbius strip has some interesting properties compared with the cylinder. Make a model of a cylinder and a Möbius strip by using strips of paper, say 40 cm by 4 cm. Draw a pencil line midway between the edges of the cylinder and of the Möbius strip. Now cut along the pencil lines. What is the result in each case? What if we cut along a line one-third of the distance between the edges?

The next result gives sufficient conditions to ensure that quotients of homeomorphic spaces are homeomorphic.

5.5   Theorem

Let $f\colon X \to Y$ be a function between the topological spaces X and Y. Suppose that X and Y have equivalence relations $\sim_X$ and $\sim_Y$ respectively such that $x \sim_X x'$ if and only if $f(x) \sim_Y f(x')$. If f is a homeomorphism then $X/\!\sim_X$ and $Y/\!\sim_Y$ are homeomorphic.

Proof Define a function $F\colon X/\!\sim_X \to Y/\!\sim_Y$ by $F[x] = [f(x)]$, where the square brackets denote equivalence classes. F is well defined since if $[x] = [x']$ then $x \sim_X x'$, thus $f(x) \sim_Y f(x')$ and $[f(x)] = [f(x')]$. We shall prove that F is a homeomorphism. To show that F is injective assume that $F[x] = F[x']$ so that $[f(x)] = [f(x')]$, i.e. $f(x) \sim_Y f(x')$. But then $x \sim_X x'$ and $[x] = [x']$. Surjectivity of F is easy to show. To prove that F is continuous we consider the natural projections $\pi_X\colon X \to X/\!\sim_X$ and $\pi_Y\colon Y \to Y/\!\sim_Y$ which are continuous. Clearly $F\pi_X = \pi_Y f$ and since f is continuous we deduce that $\pi_X F$ is continuous and hence F is continuous by the universal mapping property of quotients. The fact that F^{-1} is continuous follows in a similar way because $F^{-1}\pi_Y = \pi_X f^{-1}$.

As an example consider $\mathbb{R}^+ = (0,\infty) \subseteq \mathbb{R}$ with the equivalence relation $x \sim x'$ if and only if there is an integer n such that $x' = 3^n x$. Also consider $\mathbb{R}$ with the equivalence relation $x \sim x'$ if and only if there is an integer n such that $x' = n + x$. The function f: $\mathbb{R}^+ \to \mathbb{R}$ given by $f(x) = \log_3(x)$ is a homeomorphism and $x \sim x' \Leftrightarrow f(x) \sim f(x')$, hence the spaces $\mathbb{R}^+ /\sim$ and $\mathbb{R} /\sim$ are homeomorphic; in fact both are homeomorphic to the circle.

Theorem 5.5 explains the intuitive idea of a 'homeomorphism' as presented in Chapter 4: We start with a space W. By cutting it we get X and a relation $\sim_X$ which tells us how to reglue X in order to get W. Now perform a homeomorphism f on X to give Y with an equivalence relation $\sim_Y$. Naturally, we want that

$$x \sim_X x' \Leftrightarrow f(x) \sim_Y f(x').$$

Regluing Y according to $\sim_Y$ gives us $Z = Y/\sim_Y$. By Theorem 5.5 the space Z is homeomorphic to the space X.

A concept that we shall find useful later on is that of a group G 'acting' on a set X. The notion is fruitful and leads to examples of spaces with the quotient topology.

5.6 Definition

Let X be a set and let G be a group. We say that G *acts* on X and that X is a G-*set* if there is a function from G $\times$ X to X, denoted by $(g,x) \to g \cdot x$, such that

(i) $1 \cdot x = x$ for all $x \in X$, where 1 is the identity element of G,

(ii) $g \cdot (h \cdot x) = (gh) \cdot x$ for all $x \in X$ and $g,h \in G$.

As an example let G be the group of homeomorphisms of a topological space X (see Exercise 3.7(d)) and define $g \cdot x = g(x)$ for $g \in G$. This defines an action of G on X since clearly $1 \cdot x = 1(x) = x$ and $g \cdot (h \cdot x) = g \cdot h(x) = g(h(x)) = (gh)(x) = (gh) \cdot x$. Another example is to take $G = \mathbb{Z}_2 = \{\pm 1\}$ the group of order 2 and $X = S^n$. An action of $\mathbb{Z}_2$ on S^n is given by $\pm 1 \cdot x = \pm x$, as is easily verified. If we take $G = \mathbb{Z}$ the integers and $X = \mathbb{R}$ then an action of $\mathbb{Z}$ on $\mathbb{R}$ is given by $n \cdot x = n + x$ where $n \in \mathbb{Z}, x \in \mathbb{R}$. This example can be generalized to an action of $\mathbb{Z} \times \mathbb{Z}$ on $\mathbb{R}^2$ by $(m,n) \cdot (x,y) = (m + x, n + y)$. In both cases we leave it for the reader to verify that we do indeed get an action as defined in 5.6. Our last example for the present is an action of $\mathbb{Z}$ on the infinite strip

$$\{ (x,y) \in \mathbb{R}^2 ; -\tfrac{1}{2} \leq y \leq \tfrac{1}{2} \}$$

which is given by

$$m \cdot (x,y) = (m + x, (-1)^m y).$$

Our definition of a G-action is strictly speaking that of a *left* G-action. There is also the notion of a *right* G-action where now we have a function $X \times G \to X$, denoted by $(x,g) \to x \cdot g$ such that $x \cdot 1 = x$ and $(x \cdot g) \cdot h = x \cdot (gh)$. By a G-action we shall always mean a left G-action.

5.7 Exercises

(a) Suppose that X is a right G-set. For $x \in X$ and $g \in G$ define

$$g \cdot x = x \cdot (g^{-1})$$

Show that this defines a (left) action of G on X. Why does the definition $g \cdot x = x \cdot g$ fail?

(b) Let H be a subgroup of a group G. For $h \in H$, $g \in G$ define $h \cdot g$ to be hg. Show that this defines an action of H on G.

(c) Let G be a group and let $\mathscr{S}(G)$ denote the set of subsets of G. Show that

$$g \cdot U = gU = \{ gh; h \in U \}, g \in G, U \in \mathscr{S}(G)$$

defines an action of G on $\mathscr{S}(G)$.

(d) Let G act on X and define the *stabilizer* of $x \in X$ to be the set

$$G_x = \{ g \in G; g \cdot x = x \}.$$

Prove that G_x is a subgroup of G.

(e) Let G act on X and define the *orbit* of $x \in X$ to be the subset

$$G \cdot x = \{ g \cdot x; g \in G \}$$

of X. Prove that two orbits $G \cdot x$, $G \cdot y$ are either disjoint or equal. Deduce that a G-set X decomposes into a union of disjoint subsets.

An important consequence of the definition of a G-set X is that in fact G acts on X via bijections.

5.8 Theorem

Let X be a G-set. For any $g \in G$ the function $\theta_g \colon X \to X$ defined by $x \to g \cdot x$ is bijective.

Proof From the definition of a G-set we see that $\theta_g \theta_h = \theta_{gh}$ and $\theta_1 = 1_X$; thus $\theta_g \theta_{g^{-1}} = 1_X = \theta_{g^{-1}} \theta_g$ and so θ_g is bijective.

If G acts on X then we can define an equivalence relation $\sim$ on X by

$$x \sim y \Leftrightarrow \text{there exists } g \in G \text{ such that } g \cdot x = y,$$

or in other words x ~ y if and only if y ∈ G·x; see Exercise 5.7 (e). Now denote the set of equivalence classes by X/G; this is called the *quotient set* of X by G. There is an obvious surjective mapping X → X/G. If X is a topological space upon which G acts then we can give X/G the quotient topology. We call X/G with the quotient topology the *quotient space* of X by G.

For example if $\mathbb{Z}_2$ acts on S^n by $\pm 1 \cdot x = \pm x$ then $S^n/\mathbb{Z}_2$ is just $\mathbb{R}P^n$. If $\mathbb{Z}$ acts on $\mathbb{R}$ by $n \cdot x = n + x$ then $\mathbb{R}/\mathbb{Z}$ is just S^1.

5.9   Exercises

(a) Let X be the infinite strip $\{(x,y) \in \mathbb{R}^2 ; -\frac{1}{2} \le y \le \frac{1}{2}\}$ in $\mathbb{R}^2$ with $\mathbb{Z}$ acting on it by $m \cdot (x,y) = (m + x, (-1)^m x)$. Show that the quotient space $X/\mathbb{Z}$ is homeomorphic to the Möbius strip.

(b) Let X and Y be G-sets. We say that the function f: X → Y is G-*equivariant* if $f(g \cdot x) = g \cdot f(x)$ for all $x \in X$ and all $g \in G$. Prove that if X and Y are topological spaces and f is a G-equivariant homeomorphism (i.e. both G-equivariant and a homeomorphism) then X/G and Y/G are homeomorphic.

(c) Construct examples to show that if X and Y are topological spaces with G acting on them such that $X/G \cong Y/G$ then X and Y are not necessarily homeomorphic.

(d) Let X be a G-set. For each $x \in X$ the stabilizer G_x acts on G and so the quotient G/G_x is defined. Show that G/G_x is just the set of left cosets of G_x in G. Show that there is a G-equivariant bijection between G·x the orbit of x and G/G_x.

In the examples that we gave of groups acting on topological spaces the group in question acted continuously; we have a special name for such a space.

5.10   Definition

Suppose that X is a topological space and G is a group then we say that X is a *G-space* if G acts on X and if the function θ_g given by $x \to g \cdot x$ is continuous for all $g \in G$.

5.11   Exercise

Suppose that X is a G-space. Prove that the function θ_g given by $x \to g \cdot x$ is a homeomorphism from X to itself for all $g \in G$. Deduce that there is a homomorphism from G to the group of homeomorphisms of X.

Because of the above exercise we sometimes say that if X is a G-space then G is a group of homeomorphisms of X. Using this we prove the next result.

5.12   Theorem

Suppose that X is a G-space. Then the canonical projection π: $X \to X/G$ is an open mapping.

Proof Let U be an open set in X then consider $\pi^{-1}(\pi(U))$.

$$\begin{aligned}
\pi^{-1}(\pi(U)) &= \{\, x \in X; \pi(x) \in \pi(U)\,\} \\
&= \{\, x \in X; G{\cdot}x = G{\cdot}y \text{ for some } y \in U \,\} \\
&= \{\, x \in X; x = g{\cdot}y \text{ for some } y \in U, \text{ some } g \in G \,\} \\
&= \{\, x \in X; x \in g{\cdot}U \text{ for some } g \in G \,\} \\
&= \bigcup_{g \in G} g{\cdot}U
\end{aligned}$$

The action of each g in G is a homeomorphism, so if U is open then so is $\pi^{-1}(\pi(U))$ and hence $\pi(U)$ is open in X/G.

In the next exercises the first is an extension of the universal mapping property of quotients. The second extends Theorem 5.12 in a special case.

5.13   Exercises

(a) Let X be a G-space and let π: $X \to X/G$ be the canonical projection. Suppose that g is a function from X/G to a topological space Z. Prove that g is an open mapping if and only if $g\pi$ is an open mapping.

(b) Let X be a G-space with G finite. Prove that the natural projection π: $X \to X/G$ is a closed mapping.

(c) Suppose X is a G-space and H is a normal subgroup of G. Show that X/H is a (G/H)-space and that

$$(X/H)/(G/H) \cong X/G.$$

6

___

Product spaces

Our final general method of constructing new topological spaces from old ones is through the direct product. Recall that the direct product $X \times Y$ of two sets X, Y is the set of ordered pairs (x, y) with $x \in X$ and $y \in Y$. If X and Y are topological spaces we can use the topologies on X and Y to give one on $X \times Y$. A first guess might be that the open sets of $X \times Y$ should be products of open sets in X and in Y; this however is not quite sufficient (think! – which condition for a topology fails?).

6.1   Definition
Let X and Y be topological spaces. The *(topological) product* $X \times Y$ is the set $X \times Y$ with topology $\mathcal{U}_{X \times Y}$ consisting of the family of sets that are unions of products of open sets of X, Y.

A typical element of $\mathcal{U}_{X \times Y}$ is of the form $\bigcup_{j \in J} U_j \times V_j$ where J is some indexing set and for each $j \in J$, U_j and V_j are open subsets of X and Y respectively. That $\mathcal{U}_{X \times Y}$ is a topology is not hard to check: $\emptyset = \emptyset \times \emptyset$ and $X \times Y = X \times Y$ so that the first condition is satisfied. If $W, W' \in \mathcal{U}_{X \times Y}$ then $W = \bigcup_{j \in J} U_j \times V_j$ and $W' = \bigcup_{k \in K} U'_k \times V'_k$ for some indexing sets J, K and with U_j, U'_k open in X and V_j, V'_k open in Y. Since

$$W \cap W' = \bigcup_{(j,k) \in J \times K} (U_j \cap U'_k) \times (V_j \cap V'_k)$$

we see that condition (ii) for a topology is satisfied. The third condition is trivially true.

The notion of the topological product of X and Y can be extended to the topological product of a finite number of topological spaces in an obvious way.

6.2   Exercises

(a) Show that if $X_1 \cong X_2$ and $Y_1 \cong Y_2$ then $X_1 \times Y_1 \cong X_2 \times Y_2$.

(b) Let X,Y be metrizable spaces and suppose that they arise from metrics d_X, d_Y respectively. Show that d defined by

$$d((x_1,y_1),(x_2,y_2)) = \max \{ d_X(x_1,x_2), d_Y(y_1,y_2) \}$$

is a metric on $X \times Y$ which produces the product space topology on $X \times Y$. Deduce that the product topology on $\mathbb{R}^n \times \mathbb{R}^m$ ($\mathbb{R}^n, \mathbb{R}^m$ with usual topology), is the same as the usual topology on $\mathbb{R}^{n+m} = \mathbb{R}^n \times \mathbb{R}^m$.

(c) The *graph* of a function f: $X \to Y$ is the set of points in $X \times Y$ of the form (x,f(x)) for $x \in X$. Show that if f is a continuous function between topological spaces then the graph of f is homeomorphic to X.

(d) Prove that $\mathbb{R}^2 - \{0\}$ is homeomorphic to $\mathbb{R} \times S^1$. (Hint: Consider $\mathbb{R}^2 - \{0\}$ as $\mathbb{C} - \{0\}$.)

There is another characterization of the topology on $X \times Y$.

6.3   Theorem

Let $X \times Y$ be the product of two topological spaces. A set $W \subseteq X \times Y$ is open if and only if for all $w \in W$ there exist sets U_w, V_w such that U_w is open in X, V_w is open in Y, $U_w \times V_w \subseteq W$ and $w \in U_w \times V_w$.

Proof Suppose W is open, then $W = \underset{j \in J}{\cup} U_j \times V_j$ where J is some indexing set

and U_j, V_j are open in X,Y respectively. So, if $w \in W$ then $w \in U_i \times V_i$ for some $i \in J$. Conversely the set $\underset{w \in W}{\cup} U_w \times V_w$ is open in $X \times Y$ and clearly is equal to W.

There are obvious projection maps $\pi_X: X \times Y \to X$ and $\pi_Y: X \times Y \to Y$ given by $(x,y) \to x$ and $(x,y) \to y$. These are called the product projections. Since $\pi_X^{-1}(U) = U \times Y$ and $\pi_Y^{-1}(V) = X \times V$ it is clear that both π_X and π_Y are continuous maps.

6.4   Theorem

For all $y \in Y$ the subspace $X \times \{y\} \subseteq X \times Y$ is homeomorphic to X.

Proof Consider the map f: $X \times \{y\} \to X$ given by $(x,y) \to x$. This is clearly bijective. We may write f as the composite of the inclusion $X \times \{ y \}$ $\to X \times Y$ and the projection $\pi_X: X \times Y \to X$, both of which are continuous. Thus f is continuous. Next, suppose that W is an open subset of $X \times \{ y \}$ so that $W = (\bigcup_{j \in J} U_j \times V_j) \cap X \times \{ y \}$ where U_j, V_j are open in X,Y respectively. W may be rewritten as $\bigcup_{j \in J'} U_j \times \{ y \}$ where $J' = \{ j \in J; y \in V_j \}$, thus $f(W) = \bigcup_{j \in J'} U_j$ which is open in X. This proves that f is also open and hence is a homeomorphism.

If $f: A \to X$ and $g: A \to Y$ are mappings between topological spaces then we can define a mapping h: $A \to X \times Y$ by $h(a) = (f(a),g(a))$. It is clear that h is the unique mapping such that $\pi_X h = f$ and $\pi_Y h = g$. The relationship between the continuity of h and f,g is called the *universal mapping property of products*.

6.5   Theorem
Let A,X and Y be topological spaces. Then for any pair of mappings f: $A \to X$, g: $A \to Y$ the mapping h: $A \to X \times Y$ defined by $h(a) = (f(a),g(a))$ is continuous if and only if f and g are continuous.

Proof If h is continuous then so is $\pi_X h = f$ and $\pi_Y h = g$. Conversely suppose that f and g are continuous. Let U,V be open subsets of X,Y respectively. Then $h^{-1}(U \times V) = \{ a;f(a) \in U, g(a) \in V \} = f^{-1}(U) \cap g^{-1}(V)$, but since $f^{-1}(U)$ and $g^{-1}(V)$ are both open so is $h^{-1}(U \times V)$. Now consider an open set W in $X \times Y$. If $x \in W$ then $x \in U \times V \subseteq W$ where U,V are open in X,Y. Thus $h^{-1}(x) \in f^{-1}(U) \cap g^{-1}(V) \subseteq h^{-1}(W)$ and so $h^{-1}(W)$ is open.

6.6   Exercises
(a) Show that the product topology on $X \times Y$ is the weakest topology such that π_X and π_Y are continuous.
(b) Let X be a G-space and let Y be an H-space. Prove that the space $(X \times Y)/(G \times H)$ is homeomorphic to $(X/G) \times (Y/H)$.
(c) For $(n,m) \in \mathbb{Z} \times \mathbb{Z}$ and $(x,y) \in \mathbb{R}^2$ define
$$(n,m) \cdot (x,y) = (n+x,m+y)$$
Show that this makes $\mathbb{R}^2$ into a $(\mathbb{Z} \times \mathbb{Z})$-space. Prove that $\mathbb{R}^2/(\mathbb{Z} \times \mathbb{Z})$ is homeomorphic to $S^1 \times S^1$.
(d) Prove that the torus (see Figures 5.3(a) and 5.4) is homeomorphic to $S^1 \times S^1$.

(e) For $n \in \mathbb{Z}, z \in \mathbb{C} - \{0\}$ define $n \cdot z$ by

$n \cdot z = 2^n z$.

Show that this makes $\mathbb{C} - \{0\}$ into a $\mathbb{Z}$-space. Prove that $(\mathbb{C} - \{0\})/\mathbb{Z}$ is homeomorphic to $S^1 \times S^1$. (Hint: Use Exercises 6.2(d), 6.6(b) and the fact that $\mathbb{Z} = \mathbb{Z} \times \{1\}$.)

(f) As in (e) above but define $n \cdot z$ by

$n \cdot z = (2\omega)^n z$

where $\omega = \exp(2\pi i/3)$. What is $(\mathbb{C} - \{0\})/\mathbb{Z}$?

(g) Prove that $\mathbb{R}^n - \{0\}$ and $S^{n-1} \times \mathbb{R}$ are homeomorphic spaces. (Hint: Consider the function $f: S^{n-1} \times \mathbb{R} \to \mathbb{R}^n - \{0\}$ given by $f(x,t) = 2^t x$.)

(h) Prove that the subset $S_{p,q} \subseteq \mathbb{R}^n$ defined by

$S_{p,q} = \{x \in \mathbb{R}^n; x_1^2 + x_2^2 + ... + x_p^2 - x_{p+1}^2 - ... - x_{p+q}^2 = 1\}$

where $p + q \leq n$, is homeomorphic to $S^{p-1} \times \mathbb{R}^{n-q}$. (Hint: Consider the function $f: S^{p-1} \times \mathbb{R}^{n-q} \to S_{p,q}$ given by

$f(x_1,...x_p, y_1,...,y_{n-p}) = (x_1 z, x_2 z, ..., x_p z, y_1, y_2, ..., y_{n-p})$

where $z = \sqrt{(1 + y_1^2 + y_2^2 + ... + y_q^2)}$.)

(i) Let G be the group of homeomorphisms $\{T^i; i \in \mathbb{Z}\}$ where $T: \mathbb{R}^n - \{0\} \to \mathbb{R}^n - \{0\}$ is given by $Tx = 2x$. Show that $(\mathbb{R}^n - \{0\})/G$ is homeomorphic to $S^{n-1} \times S^1$.

(j) Prove that the following two subsets of $\mathbb{R}^n$ with the usual topology are homeomorphic.

$I^n = \{x = (x_1, x_2, ..., x_n) \in \mathbb{R}^n; 0 \leq x_i \leq 1, i = 1, 2, ..., n\}$,
$D^n = \{x \in \mathbb{R}^n; \|x\| \leq 1\}$.

(Hint: First show that $I^n \cong ([-1, 1])^n = X$, then define $\varphi: X \to D^n$ and $\psi: D^n \to X$ by

$$\varphi(x_1, x_2, ..., x_n) = \frac{\max\{|x_1|, |x_2|, ..., |x_n|\}}{\|x\|} (x_1, x_2, ..., x_n),$$
$$\varphi(0) = 0,$$

$$\psi(x_1, x_2, ..., x_n) = \frac{\|x\|}{\max\{|x_1|, |x_2|, ..., |x_n|\}} (x_1, x_2, ..., x_n),$$
$$\psi(0) = 0.$$

Intuitively: Shrink down each line segment from 0 to ∂X linearly to have length 1. See Figure 6.1.)

Figure 6.1

(k) Prove that $\overset{\circ}{D}{}^n \cong \mathbb{R}^n$. (Hint: $\overset{\circ}{D}{}^n \cong \overset{\circ}{I}{}^n = (\overset{\circ}{I})^n$.)

(l) Find a non-empty space X such that $X \cong X \times X$. (Hint: Try a non-finite set with the discrete topology. Having done that try also to find an example of such a space without the discrete topology.)

7

Compact spaces

In this and the next two chapters we shall look at properties of spaces that are preserved under homeomorphisms. One important consequence of this is that if one space has the property in question and another does not, then the two spaces cannot be homeomorphic. The first property is compactness. This concept is essentially based on the fact that if $\{\, U_j; j \in J\,\}$ is a collection of open subsets of the unit interval $[0,1] \subseteq \mathbb{R}$ (with the induced topology) such that $\bigcup_{j \in J} U_j = [0,1]$ then there is a finite subcollection of these open sets whose union still gives $[0,1]$; see Theorem 7.7.

7.1 Definition

A *cover* of a subset S of a set X is a collection of subsets $\{\, U_j; j \in J\,\}$ of X such that $S \subseteq \bigcup_{j \in J} U_j$. If in addition the indexing set J is finite then $\{\, U_j; j \in J\,\}$ is said to be a *finite cover*.

For example the collection $\{\, [1/n, 1-1/n]; n \in \mathbb{N}\,\}$ is a cover of the subset $(0,1)$ of $\mathbb{R}$. Of course if $S = X$ then $\{\, U_j; j \in J\,\}$ is a cover of X if $X = \bigcup_{j \in J} U_j$. For example if $U_n = (n, n+3) \subseteq \mathbb{R}$ then $\{\, U_n; n \in \mathbb{Z}\,\}$ is a cover of $\mathbb{R}$.

7.2 Definition

Suppose that $\{\, U_j; j \in J\,\}$ and $\{\, V_k; k \in K\,\}$ are covers of the subset S of X. If for all $j \in J$ there is a $k \in K$ such that $U_j = V_k$ then we say that $\{\, U_j; j \in J\,\}$ is a *subcover* of the cover $\{\, V_k; k \in K\,\}$.

For example $\{\, V_r; r \in R\,\}$, where $V_r = (r, r+3) \subseteq \mathbb{R}$, is a cover of $\mathbb{R}$ and $\{\, U_n; n \in \mathbb{Z}\,\}$, where $U_n = (n, n+3)$, is a subcover of $\{\, V_r; r \in R\,\}$.

7.3 Definition

Suppose that X is a topological space and S is a subset. We say that the cover $\{ U_j; j \in J \}$ is an *open cover* of S if each U_j, $j \in J$, is an open subset of X.

7.4   Definition

A subset S of a topological space X is said to be *compact* if every open cover of S has a finite subcover.

In particular the topological space X is compact if every open cover of X has a finite subcover. The space $\mathbb{R}$ with its usual topology is not a compact space because $\{ (n, n+2); n \in \mathbb{Z} \}$ is an open cover of $\mathbb{R}$ with no finite subcover. A space X with the discrete topology is compact if and only if it is finite. This is because each point of X is open and so if X is infinite then the open cover consisting of the set of single points has no finite subcover. On the other hand if X is finite then there are only a finite number of open subsets. We shall shortly show that the unit interval [0,1] is a compact subset of $\mathbb{R}$.

7.5   Exercises

(a) Suppose that X has the finite complement topology. Show that X is compact. Show that each subset of X is compact.

(b) Prove that a topological space is compact if and only if whenever $\{ C_j; j \in J \}$ is a collection of closed sets with $\bigcap_{j \in J} C_j = \emptyset$ then there is a finite subcollection $\{ C_k; k \in K \}$ such that $\bigcap_{k \in K} C_k = \emptyset$.

(c) Let $\mathcal{F}$ be the topology on $\mathbb{R}$ defined by: $U \in \mathcal{F}$ if and only if for each $s \in U$ there is a $t > s$ such that $[s,t) \subseteq U$. Prove that the subset [0,1] of $(\mathbb{R}, \mathcal{F})$ is not compact.

A subset S of a topological space may be given the induced topology and so we have two concepts of compactness for S: as a subset of X and as a space in its own right. The two concepts coincide.

7.6   Theorem

A subset S of X is compact if and only if it is compact as a space given the induced topology.

Proof This is clear since the open subsets of S with the induced topology are of the form $U \cap S$ where U is an open subset of X. The reader should write down the details carefully.

Thus we could have defined S to be compact if it is a compact space in the induced topology.

The next result yields an important example of a compact space.

7.7   Theorem
The unit interval $[0,1] \subseteq \mathbb{R}$ is compact.

Proof Let $\{ U_j; j \in J \}$ be an open cover of $[0,1]$ and suppose that there is no finite subcover. This means that at least one of the intervals $[0,\frac{1}{2}]$ or $[\frac{1}{2},1]$ cannot be covered by a finite subcollection of $\{ U_j; j \in J \}$. Denote by $[a,b]$ one of those intervals, that is $[a,b]$ cannot be covered by a finite subcollection of $\{ U_j; j \in J \}$. Again at least one of the intervals $[a_1, \frac{1}{2}(a_1 + b_1)]$ or $[\frac{1}{2}(a_1 + b_1),b_1]$ cannot be covered by a finite subcollection of $\{ U_j; j \in J \}$; denote one such by $[a_2,b_2]$. Continuing in this manner we get a sequence of intervals $[a_1,b_1]$, $[a_2,b_2]$,...., $[a_n,b_n]$ ..., such that no finite subcollection of $\{ U_j; j \in J \}$ covers any of the intervals. Furthermore $b_n - a_n = 2^{-n}$ and $a_n \leq a_{n+1} \leq b_{n+1} \leq b_n$ for all n. This last condition implies that $a_m \leq b_n$ for every pair of integers m and n so that b_n is an upper bound for the set $\{ a_1,a_2,... \}$. Let a be the least upper bound of the set $\{ a_1,a_2,... \}$. Since $a \leq b_n$ for each n, a is a lower bound of $\{ b_1,b_2,... \}$. Let b be the greatest lower bound of the set $\{ b_1,b_2,... \}$. By definition we have $a_n \leq a \leq b \leq b_n$ for each n. But since $b_n - a_n = 2^{-n}$ we have $b - a \leq 2^{-n}$ for each n and so a = b.

Since $\{ U_j; j \in J \}$ covers $[0,1]$ and $a = b \in [0,1]$ we have $a \in U_j$ for some $j \in J$. Since U_j is open there is an open interval $(a-\epsilon,a+\epsilon) \subseteq U_j$ for some $\epsilon > 0$. Choose a positive integer N so that $2^{-N} < \epsilon$ and hence $b_N - a_N < \epsilon$. However $a \in [a_N,b_N]$ and $a - a_N < 2^{-N} < \epsilon$, $b - b_N < 2^{-N} < \epsilon$ so that $[a_N,b_N] \subseteq (a - \epsilon, a + \epsilon) \subseteq U_j$ which is a contradiction to $[a_N,b_N]$ not being covered by a finite subcollection of $\{ U_j; j \in J \}$.

The above argument could be extended to show that the unit n-cube $I^n = I \times I \times ... \times I \subseteq \mathbb{R}^n$ is a compact space where $I = [0,1] \subseteq \mathbb{R}$. We shall however give another proof later on.

7.8   Theorem
Let $f: X \to Y$ be a continuous map. If $S \subseteq X$ is a compact subspace, then $f(S)$ is compact.

Proof Suppose that $\{ U_j; j \in J \}$ is an open cover of $f(S)$; then $\{ f^{-1}(U_j); j \in J \}$ is an open cover of S. Since S is compact there is a finite subcover

$\{ f^{-1}(U_k); k \in K \}$, K finite. But $f(f^{-1}(U_k)) \subseteq U_k$ and so $\{ U_k; k \in K \}$ is a cover of f(S) which is a finite subcover of $\{ U_j; j \in J \}$.

7.9 Corollary

(a) Each interval $[a,b] \subseteq \mathbb{R}$ is compact.
(b) Suppose that X and Y are homeomorphic topological spaces; then X is compact if and only if Y is compact.
(c) If X is compact and Y has the quotient topology induced by a map f: X → Y then Y is compact.
(d) S^1 is compact.

The proof is obvious. Note that (b) tells us that a non-compact space cannot be homeomorphic to a compact space.

Not every subset of a compact space is compact; for example (0,1) is a subset of the compact space [0,1] which is not compact. This is easily seen by using the covering $\{ (1/n, 1 - 1/n); n \in \mathbb{N} \}$. However a closed subset of a compact space is compact.

7.10 Theorem

A closed subset of a compact space is compact.

Proof Let $\{ U_j; j \in J \}$ be an open cover of the subset $S \subseteq X$ where each U_j is an open subset of X. Since $\underset{j \in J}{\cup} U_j \supseteq S$ we see that $\{ U_j; j \in J \} \cup$

$\{ X - S \}$ is an open cover of X and as X is compact it has a finite subcover. This finite subcovering of X is of the form $\{ U_k; k \in K \}$ or $\{ U_k; k \in K \} \cup \{ X - S \}$ where K is finite. Hence $\{ U_k; k \in K \}$ is a finite subcover of $\{ U_j; j \in J \}$ which covers S.

We have investigated compactness under the induced topology and the quotient topology. We now look at the product topology.

7.11 Theorem

Let X and Y be topological spaces. Then X and Y are compact if and only if X × Y is compact.

Proof Suppose that X and Y are compact. Let $\{ W_j; j \in J \}$ be an open cover of X × Y. By definition each W_j is of the form $\underset{k \in K}{\cup} (U_{j,k} \times V_{j,k})$ where $U_{j,k}$ is open in X and $V_{j,k}$ is open in Y. Thus $\{ U_{j,k} \times V_{j,k}; j \in J, k \in K \}$ is an open cover of X × Y. For each $x \in X$ the subspace $\{ x \} \times Y$ is compact (it is homeomorphic to Y) and since $\{\ U_{j,k} \times V_{j,k}; j \in J, k \in K\ \}$ also covers $\{\ x\ \} \times Y$ there is a finite subcover

$$\{\ U_i(x) \times V_i(x); i = 1,2,...,n(x)\ \}$$

covering $\{\ x\ \} \times Y$. Let $U'(x)$ be defined by

$$U'(x) = \overset{n(x)}{\underset{i=1}{\cap}}\ U_i(x)$$

The collection $\{\ U'(x); x \in X\ \}$ is an open cover of X and therefore has a finite subcover $\{\ U'(x_i); i = 1,2,...,m\ \}$. Clearly

$$\{\ U'(x_i) \times V_{k_i}(x_i); i = 1,2,...,m, k_i = 1,2,...,n(x_i)\ \}$$

is a finite open cover of $X \times Y$. For each i and k_i there is some $j \in J$ and $k \in K$ such that

$$U'(x_i) \times V_{k_i}(x_i) \subseteq U_{j,k} \times V_{j,k} \subseteq W_j.$$

It follows that there is a finite subcover of $\{\ W_j; j \in J\ \}$ which covers $X \times Y$.

Conversely if $X \times Y$ is compact then X and Y are compact because π_X and π_Y are continuous.

More generally, of course, if $X_1, X_2,..., X_n$ are compact topological spaces then the product $X_1 \times X_2 \times ... \times X_n$ is also compact. In particular the unit n-cube I^n is compact. A subset S of $\mathbb{R}^n$ is said to be *bounded* if there is a real number $K > 0$ such that for each point $x = (x_1, x_2,..., x_n) \in S, |x_i| \leq K$ for $i = 1,2,...,n$. In other words S lies inside the n-cube of width $2K$. Since this is homeomorphic to the unit n-cube we deduce:

7.12   Theorem

(Heine–Borel) A closed and bounded subset of $\mathbb{R}^n$ is compact.

The converse to Theorem 7.12 is also true; see Exercise 8.14(n). From previous results we may now deduce that each of the following spaces is compact:

S^n (closed bounded subset of $\mathbb{R}^{n+1}$);

$S^n \times S^n \times ... \times S^n$;

$\mathbb{R}P^n$ (surjective image of S^n);

the Möbius strip (closed bounded subset of $\mathbb{R}^3$).

7.13   Exercises

(a) Which of the following spaces are compact?

$D^n = \{\ x \in \mathbb{R}^n; \|x\| \leq 1\ \}, \{\ x \in \mathbb{R}^n; \|x\| < 1\ \},$

$\{\ (s,t) \in \mathbb{R}^2; 0 \leq s \leq 1, 0 \leq t \leq 4\ \},$

$\{ (s,t,u) \in \mathbb{R}^3 ; s^2 + t^2 \leq 1 \} \cap \{ (s,t,u) \in \mathbb{R}^3 ; t^2 + u^2 \leq 1 \}$

(b) Prove that a compact subset of $\mathbb{R}^n$ is bounded.

(c) Prove that the graph of the function f: $I \to \mathbb{R}$ is compact if and only if f is continuous. Give an example of a discontinuous function g: $I \to \mathbb{R}$ with a graph which is closed but not compact.

(d) Let X,Y be topological spaces. Let $\mathscr{F}(X,Y)$ be the set of continuous functions from X to Y. If $A \subseteq X$ and $B \subseteq Y$ then write F(A,B) for the subset of $\mathscr{F}(X,Y)$ which maps A into B:

F(A,B) = $\{ f \in \mathscr{F}(X,Y); f(A) \subseteq B \}$.

Let $\mathscr{S}$ be the following

$\mathscr{S}$ = $\{ F(A,B); A$ is a compact subset of X and B is an open set of Y$\}$.

Define $\mathscr{U}$ by

$\mathscr{U}$ = $\{ U \subseteq \mathscr{F}(X,Y);$ if $f \in U$ then there are elements $F_1, F_2, ...,$ $F_n \in \mathscr{S}$ such that $f \in F_1 \cap F_2 \cap ... \cap F_n \subseteq U\}$

Prove that $\mathscr{U}$ is a topology for $\mathscr{F}(X,Y)$. (It is called the *compact-open* topology.)

(e) Let X be a compact metrizable topological space. Suppose that Y is a metric space with metric d, define d^* on $\mathscr{F}(X,Y)$ by

$d^*(f,g) = \sup_{x \in X} \ d(f(x),g(x))$.

Show that d^* is a metric for $\mathscr{F}(X,Y)$ and that the resulting topology on $\mathscr{F}(X,Y)$ is the compact-open topology.

(f) A space X is said to be *locally compact* if for all $x \in X$ every neighbourhood of x contains a compact neighbourhood of x. Show that if X is locally compact then the *evaluation* map e: $\mathscr{F}(X,Y) \times X \to Y$ given by e(f,x) = f(x) is continuous.

(g) Let X be a compact topological space arising from some metric space with metric d. Prove that if $\{ U_j; j \in J \}$ is an open cover of X then there exists a real number $\delta > 0$ (called the *Lebesgue number* of $\{ U_j; j \in J \}$) such that any subset of X of diameter less than δ is contained in one of the sets $U_j, j \in J$.

(h) Let X be a topological space and define X^∞ to be $X \cup \{ \infty \}$ where ∞ is an element not contained in X. If $\mathscr{U}$ is the topology for X then define $\mathscr{U}^\infty$ to be $\mathscr{U}$ together with all sets of the form $V \cup \{ \infty \}$ where $V \subseteq X$ and $X-V$ is both compact and closed in X. Prove that $\mathscr{U}^\infty$ is a topology for X^∞. Prove also that X is a subspace of X^∞ and X^∞ is compact. (X^∞ is called the *one-point compactification* of X.)

8

Hausdorff spaces

The starting point of this chapter is Exercise 2.2(b) where you were asked to prove that if a topological space X is metrizable then for every pair x,y of distinct points of X there are open sets U_x and U_y containing x and y respectively such that $U_x \cap U_y = \emptyset$. The proof is straightforward: since $x \neq y$, $d(x,y) = 2\epsilon$ for some ϵ where d is some metric on X realizing the topology on X. The sets $B_\epsilon(x) = \{\, z \in X; d(x,z) < \epsilon \,\}$ and $B_\epsilon(y)$ satisfy the required conditions.

8.1   Definition
        A space X is *Hausdorff* if for every pair of distinct points x,y there are open sets U_x, U_y containing x,y respectively such that $U_x \cap U_y = \emptyset$.

Thus all metrizable spaces are Hausdorff, in particular $\mathbb{R}^n$ with the usual topology and any space with the discrete topology is Hausdorff. A space with the concrete topology is not Hausdorff if it has at least two points.

8.2   Exercises
    (a) Let X be a space with the finite complement topology. Prove that X is Hausdorff if and only if X is finite.
    (b) Let $\mathscr{F}$ be the topology on $\mathbb{R}$ defined by $U \in \mathscr{F}$ if and only if for each $s \in U$ there is a $t > s$ such that $[s,t) \subseteq U$. Prove that $(\mathbb{R}, \mathscr{F})$ is Hausdorff.
    (c) Suppose that X and Y are homeomorphic topological spaces. Prove that X is Hausdorff if and only if Y is Hausdorff.

The Hausdorff condition is an example of a *separation* condition. We shall define some of the other separation conditions, but apart from the next few pages we shall only pursue the Hausdorff condition in detail.

8.3   Definition
        Let k be one of the integers 0,1,2,3 or 4. A space X is said to be a

T_k-*space* if it satisfies condition T_k given below:

T_0: For every pair of distinct points there is an open set containing one of them but not the other.

T_1: For every pair x,y of distinct points there are two open sets, one containing x but not y, and the other containing y but not x.

T_2: For every pair x,y of distinct points there are two disjoint open sets, one containing x and the other containing y.

T_3: X satisfies T_1 and for every closed subset F and every point x not in F there are two disjoint open sets, one containing F and the other containing x.

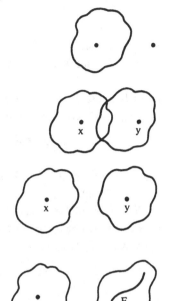

T_4: X satisfies T_1 and for every pair F_1, F_2 of disjoint closed subsets there are two disjoint open sets, one containing F_1 and the other containing F_2.

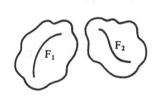

A T_2-space is a Hausdorff space. A T_3-space is sometimes called a regular space.

It is clear that $T_2 \Rightarrow T_1 \Rightarrow T_0$. The reason why condition T_1 was included in conditions T_3 and T_4 will be revealed in the next result (Theorem 8.5) from which it will follow that $T_4 \Rightarrow T_3 \Rightarrow T_2 \Rightarrow T_1 \Rightarrow T_0$.

8.4 Exercises

(a) Suppose that X and Y are homeomorphic spaces. Prove that X is a T_k-space if and only if Y is a T_k-space (k = 0,1,2,3,4).

(b) Construct topological spaces X_0, X_1, X_2 and X_3 with the property that X_k is a T_k-space but X_k is not a T_j-space for $j > k$.

(c) Prove that a compact Hausdorff space is a T_4-space. (Hint: Look at the proof of Theorem 8.7 and in desperation look at the proof of Theorem 8.11.)

8.5   Theorem
A space X is T_1 if and only if each point of X is closed.

Proof Suppose X is a T_1-space. Let $x \in X$ and $y \in X - \{ x \}$. Then there is an open set U_y containing y but not x. Therefore

$$\underset{y \in X - \{ x \}}{\cup} U_y = X - \{ x \}$$

which shows that $X - \{ x \}$ is a union of open sets and hence is open. Thus $\{ x \}$ is closed.

Conversely if $\{ x \}$ and $\{ y \}$ are closed then $X - \{ x \}$ and $X - \{ y \}$ are open sets, one containing x but not y and the other containing y but not x; i.e. X is a T_1-space.

As a corollary we get the following result.

8.6   Corollary
In a Hausdorff space each point is a closed subset.

In fact something much more general holds.

8.7   Theorem
A compact subset A of a Hausdorff space X is closed.

Proof We may suppose that $A \neq \emptyset$ and $A \neq X$ since otherwise it is already closed and there is nothing to prove. Choose a point $x \in X - A$. For each $a \in A$ there is a pair of disjoint open sets U_a, V_a with x in U_a and a in V_a. The set $\{ V_a; a \in A \}$ covers A and since A is compact there is a finite subcover, say

$$\{ V_{a(1)}, V_{a(2)}, ..., V_{a(n)} \}$$

which covers A. The set $U = U_{a(1)} \cap U_{a(2)} \cap ... \cap U_{a(n)}$ is an open set containing x which is disjoint from each of the $V_{a(i)}$ and hence $U \subseteq X - A$. Thus each point $x \in X - A$ has an open set containing it which is contained in $X - A$, which means that $X - A$ is open and A is closed.

Theorem 8.7 leads to an important result.

8.8   Theorem
Suppose that f: $X \to Y$ is a continuous map from a compact space X

to a Hausdorff space Y. Then f is a homeomorphism if and only if f is bijective.

Proof Obviously if f is a homeomorphism then f is bijective. It is the converse which is more interesting. Suppose therefore that f is bijective, whence f^{-1} exists. Now f^{-1} is continuous if and only if $(f^{-1})^{-1}(V) = f(V)$ is closed whenever V is closed in X. If V is closed in X then V is compact by 7.10, whence f(V) is compact by 7.8 and so f(V) is closed by 8.7 which proves that f^{-1} is continuous.

We need both the Hausdorff and compactness conditions in the above result. For example if X is the real numbers with the discrete topology (hence not compact) and Y is $\mathbb{R}$ with the usual topology (hence Hausdorff) then the identity map $X \to Y$ is continuous and bijective but not a homeomorphism. Also, if $X = \{$ x,y $\}$ with the discrete topology (hence compact) and $Y = \{$ x,y $\}$ with the topology $\{$ $\emptyset$,Y, $\{$ x $\}\}$ (hence not Hausdorff) then the identity map is continuous and bijective but not a homeomorphism.

Using the above theorem, many of the homeomorphisms in Chapter 5 can now be easily seen. For example the image f(X) of a compact space X in a Hausdorff space under a continuous injective map is homeomorphic to X.

We now go on to investigate how the Hausdorff property carries over to subspaces, topological products and quotient spaces.

8.9 Theorem

A subspace S of a Hausdorff space X is Hausdorff.

Proof Let x,y be a pair of distinct points in S. Then there are a pair of disjoint open sets U_x and U_y in X with x in U_x and y in U_y. The sets $(U_x \cap S)$ and $(U_y \cap S)$ are disjoint open sets in S and x is in $(U_x \cap S)$ while y is in $(U_y \cap S)$. Hence S is Hausdorff.

In particular every subset of $\mathbb{R}^n$ with the usual topology is Hausdorff.

8.10 Theorem

Let X and Y be topological spaces. Then X and Y are Hausdorff if and only if $X \times Y$ is Hausdorff.

Proof Suppose that X and Y are Hausdorff and let $w_1 = (x_1,y_1)$ and $w_2 = (x_2,y_2)$ be two distinct points of $X \times Y$. If $x_1 \neq x_2$ then we can find two open disjoint sets U_1,U_2 with $x_1 \in U_1$, $x_2 \in U_2$. The sets $U_1 \times Y$ and $U_2 \times Y$ are disjoint open sets in $X \times Y$ with $w_1 \in U_1 \times Y$ and $w_2 \in U_2 \times Y$.

If $x_1 = x_2$ then $y_1 \neq y_2$ and a similar argument shows that there are disjoint open sets $X \times V_1$ and $X \times V_2$ in $X \times Y$ with $w_1 \in X \times V_1$ and $w_2 \in X \times V_2$.

Conversely, if $X \times Y$ is Hausdorff then so are the subspaces $X \times \{ y \}$, $\{ x \} \times Y$ and hence so are X and Y.

Thus spaces like $S^1 \times S^1 \times ... \times S^1$ are Hausdorff.

Although subspaces of Hausdorff spaces are Hausdorff and products of Hausdorff spaces are Hausdorff it is not true in general that a quotient space of a Hausdorff space is Hausdorff. As an example let X be a Hausdorff space with a subset A of X which is not closed, (e.g. $X = \mathbb{R}$, $A = (0,1)$). Let Y be $X/\sim$ where $\sim$ is the equivalence relation on X given by $x \sim x'$ if and only if $x = x'$ or $\{ x,x' \} \subseteq A$ (intuitively Y is X with A shrunk to a point; in general we denote $Y = X/\sim$ by X/A). If we give Y the quotient topology with g: $X \to Y$ being the natural projection then the inverse image of the point $[x_0] \in Y$ where $x_0 \in A$ is A which is not closed in X. Therefore the point $[x_0]$ is not closed in Y and Y is not Hausdorff.

To ensure that a quotient space Y of a Hausdorff space X is Hausdorff we need to impose further restrictions on X. As an example we give the following result.

8.11   Theorem

Let Y be the quotient space of the topological space X determined by the surjective mapping f: $X \to Y$. If X is compact Hausdorff and f is closed then Y is (compact) Hausdorff.

Proof Points of Y are images of points in X which are closed in X, thus the points of Y are closed. Let y_1 and y_2 be a pair of distinct points in Y. The sets $f^{-1}(y_1)$ and $f^{-1}(y_2)$ are disjoint closed subsets of X. For each point $x \in f^{-1}(y_1)$ and each point $a \in f^{-1}(y_2)$ there are a pair of disjoint open sets $U_{x,a}$ and $V_{x,a}$ with $x \in U_{x,a}$ and $a \in V_{x,a}$. Since $f^{-1}(y_2)$ is closed it is also compact and so there is a finite subcover of $\{ V_{x,a}; a \in f^{-1}(y_2) \}$ which covers $f^{-1}(y_2)$, say $\{ V_{x,a}; a \in A \}$ where A is a finite subset of $f^{-1}(y_2)$. In particular there are disjoint open sets U_x and V_x with $x \in U_x$ and $f^{-1}(y_2) \subseteq V_x$, in fact

$$U_x = \bigcap_{a \in A} U_{x,a}, \quad V_x = \bigcup_{a \in A} V_{x,a}.$$

Now $\{ U_x; x \in f^{-1}(y_1) \}$ is an open cover of $f^{-1}(y_1)$ which is compact, hence there is a finite subcover $\{ U_x; x \in B \}$ where B is a finite subset of $f^{-1}(y_1)$. Thus the sets

$$U = \bigcup_{x \in B} U_x, \quad V = \bigcap_{x \in B} V_x$$

are disjoint open sets with $f^{-1}(y_1) \subseteq U$ and $f^{-1}(y_2) \subseteq V$.

Since f is closed, by assumption, $f(X - U)$ and $f(X - V)$ are closed subsets of Y so that $W_1 = Y - f(X - U)$ and $W_2 = Y - f(X - V)$ are open subsets of Y with $y_1 \in W_1$ and $y_2 \in W_2$. Finally, we just need to check that $W_1 \cap W_2 = \emptyset$. Suppose therefore that $y \in W_1 \cap W_2$, then $y \notin f(X - U)$ and $y \notin f(X - V)$. Therefore $f^{-1}(y) \cap (X - U) = \emptyset$ and $f^{-1}(y) \cap (X - V) = \emptyset$ from which it follows that $f^{-1}(y) \subseteq U \cap V = \emptyset$ and hence $W_1 \cap W_2 = \emptyset$.

As a corollary we get the following result.

8.12 Corollary

If X is a compact Hausdorff G-space with G finite then X/G is a compact Hausdorff space.

Proof Let C be a closed subset of X. Then

$$\pi^{-1}(\pi(C)) = \bigcup_{g \in G} g \cdot C$$

where $\pi: X \to X/G$ is the natural projection. Since the action of $g \in G$ on X is a homeomorphism $g \cdot C$ is closed for all $g \in G$. Thus $\pi^{-1}(\pi(C))$ is closed and hence $\pi(C)$ is closed which shows that π is a closed mapping.

So, for example, $\mathbb{R} P^n$ is a compact Hausdorff space.

For another corollary to Theorem 8.11 consider a space X with a subset $A \subseteq X$. Recall that X/A denotes $X/\sim$ where $\sim$ is the equivalence relation on X given by

$$x \sim x' \text{ if and only if } x = x' \text{ or } x, x' \in A$$

8.13 Corollary

If X is a compact Hausdorff space and A is a closed subset of X then X/A is a compact Hausdorff space.

Proof Let C be a closed subset of X and let p: $X \to X/A$ denote the natural map. If $C \cap A = \emptyset$ then $p(C) = C$ is closed. If $C \cap A \neq \emptyset$ then $p(C) = p(C - A) \cup p(C \cap A)$ which is closed because $p^{-1}(p(C - A) \cup p(C \cap A)) = (C - A) \cup A = C \cup A$. Thus p is a closed map.

Other restrictions on a Hausdorff space ensuring that a quotient is also Hausdorff appear in the next set of exercises, where a converse to Theorem 8.11 is also given.

8.14   Exercises

(a) Let $f: X \to Y$ be a continuous surjective map of a compact space X onto a Hausdorff space Y. Prove that a subset U of Y is open if and only if $f^{-1}(U)$ is open in X. (Hint: Prove that a subset C of Y is closed if and only if $f^{-1}(C)$ is closed in X.) Deduce that Y has the quotient topology determined by f.

(b) Prove that the space Y is Hausdorff if and only if the diagonal $D = \{ (y_1,y_2) \in Y \times Y; y_1 = y_2 \}$ in $Y \times Y$ is a closed subset of $Y \times Y$.

(c) Let $f: X \to Y$ be a continuous map. Prove that if Y is Hausdorff then the set $\{ (x_1,x_2) \in X \times X; f(x_1) = f(x_2) \}$ is a closed subset of $X \times X$.

(d) Let $f: X \to Y$ be a map which is continuous, open and onto. Prove that Y is a Hausdorff space if and only if the set $\{ (x_1,x_2) \in X \times X; f(x_1) = f(x_2) \}$ is a closed subset of $X \times X$.

(e) Let X be a compact Hausdorff space and let Y be a quotient space determined by a map $f: X \to Y$. Prove that Y is Hausdorff if and only if f is a closed map. Furthermore, prove that Y is Hausdorff if and only if the set $\{ (x_1,x_2) \in X \times X; f(x_1) = f(x_2) \}$ is a closed subset of $X \times X$.

(f) Let $\sim$ be the equivalence relation on $S^1 \times I$ given by $(x,t) \sim (y,s)$ if and only if $xt = ys$ (here we think of $S^1 \subseteq \mathbb{C}$ and $I = [0,1] \subseteq \mathbb{R}$). Prove that $(S^1 \times I)/\sim$ is homeomorphic to the unit disc $D^2 = \{ x \in \mathbb{R}^2; \|x\| \le 1 \} = \{ x \in \mathbb{C}; |x| \le 1 \}$ with the induced topology.

(g) Let $\sim$ be the equivalence relation on the unit square region $X = \{ (x,y) \in \mathbb{R}^2; 0 \le x,y \le 1 \}$ given by $(x,y) \sim (x',y')$ if and only if $(x,y) = (x',y')$ or $\{ x,x' \} = \{ 1,0 \}$ and $y = 1-y'$ or $\{ y,y' \} = \{ 1,0 \}$ and $x = 1-x'$. See Figure 8.1. Prove that the identification space $X/\sim$ is homeomorphic to $\mathbb{R}P^2$.

Figure 8.1

(h) Let S^n_+ be the subset of $S^n \subseteq \mathbb{R}^{n+1}$ given by
$$S^n_+ = \{ x = (x_1,x_2,...,x_{n+1}) \in \mathbb{R}^{n+1}; \|x\| = 1, x_{n+1} \ge 0 \} .$$

Prove that the function f: $\mathbb{R}^{n+1} \to \mathbb{R}^n$ given by

$$f(x_1, x_2, ..., x_{n+1}) = (x_1, x_2, ..., x_n)$$

induces a homeomorphism from S_+^n to the closed n-disc D^n,

$D^n = \{ x \in \mathbb{R}^n; \|x\| \leq 1 \}$.

(i) Define $\sim$ on $\mathbb{R}$ by

x $\sim$ y if and only if x-y is rational.

Show that $\sim$ is an equivalence relation and that $\mathbb{R}/\sim$ with the quotient topology is not Hausdorff.

(j) Let X be a compact Hausdorff space and let U be an open subset of X not equal to X itself. Prove that

$U^\infty \cong X/(X-U)$.

(Hint: Consider h: $U^\infty \to X/(X-U)$ given by h(u) = p(u) for u $\in$ U and h(∞) = p(X-U) where p: X $\to$ X/(X-U) is the natural projection.) Deduce that if x $\in$ X (and X is a compact Hausdorff space) then

$(X - \{ x \})^\infty \cong X$.

(k) Prove that

$S^n \cong (\mathbb{R}^n)^\infty \cong D^n/S^{n-1} \cong I^n/\partial I^n$.

(Hint: $S^n - \{ (0,0,...,0,1) \} \cong \mathbb{R}^n \cong D^n-S^{n-1} \cong I^n-\partial I^n$.)

(l) (Generalization of 8.11) Let Y be the quotient space of X determined by the surjective mapping f: X $\to$ Y. Suppose that X is a Hausdorff space, f is a closed mapping and $f^{-1}(y)$ is compact for all y $\in$ Y. Prove that Y is a Hausdorff space.

(m) Suppose that X is a compact Hausdorff space and that A is a closed subspace of X. Suppose furthermore that A is a G-space with G finite. Define a relation $\sim$ on X by saying that x $\sim$ x' if and only if either x = x' or both x,x' $\in$ A and x = g·x' for some g $\in$ G. Prove that $\sim$ is an equivalence relation on X and prove that the space X/$\sim$ is Hausdorff.

(n) Prove that

A subset of $\mathbb{R}^n$ is compact if and only if it is closed and bounded.

(Hint: Use Theorem 7.12, Exercise 7.13(b) and Theorem 8.7.)

Connected spaces

Intuitively a space X is connected if it is in 'one piece'; but how should a 'piece' be interpreted topologically? It is reasonable to require that open or closed subsets of a 'piece' are open or closed respectively in the whole space X. Thus by Lemma 4.4 we should expect that a 'piece' is an open and closed subset of X. This leads us to the following definition.

9.1 Definition

A topological space X is *connected* if the only subsets of X which are both open and closed are $\emptyset$ and X. A subset of X is connected if it is connected as a space with the induced topology.

An equivalent definition is that X is connected if it is not the union of two disjoint non-empty open subsets of X. That this is so forms the next result.

9.2 Theorem

A space X is connected if and only if X is not the union of two disjoint non-empty open subsets of X.

Proof Let X be connected and suppose $X = X_1 \cup X_2$ where X_1 and X_2 are disjoint open subsets of X. Then $X - X_1 = X_2$ and so X_1 is both open and closed which means that $X_1 = \emptyset$ or X and $X_2 = X$ or $\emptyset$ respectively. In either case X is not the union of two disjoint non-empty open subsets of X.

Conversely, suppose that X is not the union of two disjoint non-empty open subsets of X and let $U \subseteq X$. If U is both open and closed then $X - U$ is both open and closed. But since X is then the disjoint union of the open sets U and $X - U$ one of these must be empty, i.e. $U = \emptyset$ or $U = X$.

As an example, the subset $S^0 = \{ \pm 1 \}$ of R is not connected because $\{ + 1 \}$ is both an open and a closed subset of S^0; or equivalently because S^0 is the disjoint union of the open subsets $\{ + 1 \}$ and $\{ - 1 \}$ of S^0. An example of a connected subset of R is [a,b], but this is a theorem. Before proving this let us have some more examples. The examples show that we have to be careful with our intuition.

Let X be the real numbers with the topology $\{ \emptyset \} \cup \{ R \} \cup \{ (-\infty, x); x \in R \}$; then any subset of X is connected. To prove this let S be any subset of X. Suppose that F is a non-empty subset of S which is both open and closed in S. Thus we may write F as $U \cap S = C \cap S$ where U is open in X and C is closed in X, i.e. $U = (-\infty, b)$ for some b and $C = [a, \infty)$ for some a. Since $F = U \cap S = C \cap S$ it follows that if $x \in S$ then $x < b$ and $x \geq a$ (if there is an $x \geq b$ then $C \cap S \neq U \cap S$; similarly if there is an $x < a$ then $U \cap S \neq C \cap S$). Thus $S \subseteq [a,b)$ and $F = S$ which means that S is connected.

Now let X be the real numbers with the topology $\mathscr{F}$ defined by: $S \in \mathscr{F}$ if and only if for each $s \in S$ there is a $t > s$ such that $[s,t) \subseteq S$. In this case the only non-empty connected subsets of X are single points. To prove this suppose that T is a non-empty connected subset of X and let x be a point in T. The subset $[x, x + \epsilon)$ of X is both an open and closed set (Exercise 2.6(d)) for all $\epsilon > 0$. Thus $[x, x + \epsilon) \cap T$ is an open and closed subset of T. Since T is connected and $[x, x-\epsilon) \cap T \neq \emptyset$ it follows that $[x, x + \epsilon) \cap T = T$ for all $\epsilon > 0$. But this is possible only if $T = \{ x \}$. Clearly single points are connected and so the only non-empty connected subsets of X are single points.

We come now to the proof that the subset [a,b] of R (with the usual topology) is connected.

9.3 Theorem
The interval $[a,b] \subseteq R$ is connected.

Proof Suppose that [a,b] is the disjoint union of two open sets U, V of [a,b]. Also, suppose that $a \in U$. Note that U and V are also closed in [a,b] and hence, since [a,b] is closed in R, they are also closed in R. Let h be the least upper bound of the set

$$\{ u \in U; u < v \text{ for all } v \in V \}$$

(this set is non-empty since a belongs to it). Because U is closed, $h \in U$. Now $(h-\epsilon, h+\epsilon) \cap V \neq \emptyset$ for all $\epsilon > 0$ (otherwise h would not be an upper bound) and so by Lemma 2.7, $h \in \bar{V}$. But V is closed so $h \in V$, and $h \in U \cap V$ gives a contradiction, proving that [a,b] is connected.

9.4 Theorem
The image of a connected space under a continuous mapping is connected.

Proof Suppose that X is connected and f: $X \to Y$ is a continuous surjective map. If U is open and closed in Y then $f^{-1}(U)$ is open and closed in X which means that $f^{-1}(U) = \emptyset$ or X and $U = \emptyset$ or Y. Thus Y is connected.

9.5   Corollary
If X and Y are homeomorphic topological spaces then X is connected if and only if Y is connected.

From Theorem 9.4 we deduce that the circle S^1 is connected since there is a continuous surjective map f: $[0,1] \rightarrow S^1$ given by $f(t) = (\cos(2\pi t), \sin(2\pi t)) \in S^1 \subseteq \mathbb{R}^2$.

To prove that intervals in $\mathbb{R}$ of the form [a,b), (a,b] and (a,b) are connected we make use of the next result.

9.6   Theorem
Suppose that $\{ Y_j; j \in J \}$ is a collection of connected subsets of a space X. If $\cap_{j \in J} Y_j \neq \emptyset$ then $Y = \cup_{j \in J} Y_j$ is connected.

Proof Suppose that U is a non-empty open and closed subset of Y. Then $U \cap Y_i \neq \emptyset$ for some $i \in J$ and $U \cap Y_i$ is both open and closed in Y_i. But Y_i is connected so $U \cap Y_i = Y_i$ and hence $Y_i \subseteq U$. The set Y_i intersects every other Y_j, $j \in J$ and so U also intersects every Y_j, $j \in J$. By repeating the argument we deduce that $Y_j \subseteq U$ for all $j \in J$ and hence $U = Y$.

That the subsets [a,b), (a,b] and (a,b) of $\mathbb{R}$ are connected follows from Theorem 9.3, Corollary 9.5 and the fact that

$$[a,b) = \cup_{n \geq 1} [a,b - (b-a)/2^n]$$

etc. Similarly it follows that $\mathbb{R}$ itself and intervals of the form $[a,\infty)$, $(-\infty,b]$, $(-\infty,b)$, (a,∞) are connected.

The final result that we shall prove concerns products of connected spaces.

9.7   Theorem
Let X and Y be topological spaces. Then X and Y are connected if and only if $X \times Y$ is connected.

Proof Suppose that X and Y are connected. Since $X \cong X \times \{ y \}$ and $Y \cong \{ x \} \times Y$ for all $x \in X$, $y \in Y$ we see that $X \times \{ y \}$ and $\{ x \} \times Y$ are connected. Now $(X \times \{ y \}) \cap (\{ x \} \times Y) = \{ (x,y) \} \neq \emptyset$ and so $(X \times \{ y \}) \cup (\{ x \} \times Y)$ is connected by Theorem 9.6. We may write $X \times Y$ as

$$X \times Y = \bigcup_{x \in X} ((X \times \{y\}) \cup (\{x\} \times Y))$$

for some fixed $y \in Y$. Since $\bigcap_{x \in X} ((X \times \{y\}) \cup (\{x\} \times Y)) \neq \emptyset$ we deduce

that $X \times Y$ is connected.

Conversely, suppose that $X \times Y$ is connected. That X and Y are connected follows from Theorem 9.4 and the fact that $\pi_X : X \times Y \to X$ and $\pi_Y : X \times Y \to Y$ are continuous surjective maps.

From the above results we see that $\mathbb{R}^n$ is connected. In the exercises we shall see that S^n is connected for $n \geq 1$ and also that $\mathbb{R}P^n$ is connected.

9.8 Exercises

(a) Prove that the set of rational numbers $\mathbb{Q} \subseteq \mathbb{R}$ is not a connected set. What are the connected subsets of $\mathbb{Q}$?

(b) Prove that a subset of $\mathbb{R}$ is connected if and only if it is an interval or a single point. (A subset of $\mathbb{R}$ is called an *interval* if A contains at least two distinct points, and if $a, b \in A$ with $a < b$ and $a < x < b$ then $x \in A$.)

(c) Let X be a set with at least two elements. Prove
(i) If X is given the discrete topology then the only connected subsets of X are single point subsets.
(ii) If X is given the concrete topology then every subset of X is connected.

(d) Which of the following subsets of $\mathbb{R}^2$ are connected?
$$\{x; \|x\| < 1\}, \{x; \|x\| > 1\}, \{x; \|x\| \neq 1\}.$$
Which of the following subsets of $\mathbb{R}^3$ are connected?
$$\{x; x_1^2 + x_2^2 - x_3^2 = 1\}, \{x; x_1^2 + x_2^2 + x_3^2 = -1\},$$
$$\{x; x_1 \neq 1\}.$$

(e) Prove that a topological space X is connected if and only if each continuous mapping of X into a discrete space (with at least two points) is a constant mapping.

(f) A is a connected subspace of X and $A \subseteq Y \subseteq \bar{A}$. Prove that Y is connected.

(g) Suppose that Y_0 and $\{Y_j ; j \in J\}$ are connected subsets of a space X. Prove that if $Y_0 \cap Y_j \neq \emptyset$ for all $j \in J$ then $Y = Y_0 \cup (\bigcup_{j \in J} Y_j)$ is connected.

(h) Prove that $\mathbb{R}^{n+1} - \{0\}$ is connected if $n \geq 1$. Deduce that S^n and

$R P^n$ are connected for $n \geq 1$. (Hint: Consider f: $R^{n+1} - \{0\} \rightarrow S^n$ given by $f(x) = x/\|x\|$.)

(i) Let A and B be subsets of R^2 defined by

$A = \{ (x,y); x = 0, -1 \leq y \leq 1 \}$,
$B = \{ (x,y); 0 < x \leq 1, y = \cos(\pi/x) \}$.

Prove that $X = A \cup B$ is connected. (Hint: Prove that A and B are connected. Then consider $X = U \cup V$ where U,V are open and closed in X. Finally assume that some point of A is in U.)

(j) Let A and B be subsets of R^2 defined by

$A = \{ (x,y); \frac{1}{2} \leq x \leq 1, y = 0 \}$,
$B = \{ (x,y); 0 \leq x \leq 1, y = x/n$ where $n \in N \}$.

Prove that $X = A \cup B$ is connected.

(k) *First steps in algebraic topology.* Let X be a topological space and define H(X) to be the set of continuous maps from X to Z_2 (the topological space consisting of two points $\{ 0,1 \}$ with the discrete topology). If f, g $\in$ H(X) then define f + g by

$(f + g)(x) = f(x) + g(x) \bmod 2 \ (x \in X)$.

Prove that f + g is continuous and H(X) is an abelian group with respect to this operation. Prove that X is connected if and only if H(X) is isomorphic to the cyclic group of order 2. Construct examples of topological spaces X_k with $H(X_k)$ isomorphic to $(Z_2)^k$.

10

The pancake problems

In this chapter we give some light hearted applications of the results of previous chapters by looking at the so-called 'pancake problems'. Roughly stated the first problem is: Suppose you have two pancakes (of any shape) on a plate, show that it is possible to cut both exactly in half with just one stroke of a knife. The second problem is to show that you can divide one pancake into four equal parts with two perpendicular cuts of a knife. The proofs are based on a form of the *intermediate value theorem*.

10.1 Lemma

If $f: I \to \mathbb{R}$ is a continuous function such that the product $f(0)$ $f(1)$ is finite and non-positive then there exists a point $t \in I$ such that $f(t) = 0$.

Proof Suppose that $f(t) \neq 0$ for all $t \in I$; in particular $f(0)\, f(1) < 0$. Define a function $g: I \to \{\pm 1\} = S^0$ by $g(t) = f(t)/(|f(t)|)$. This is clearly continuous and surjective (because $f(0)\, f(1) < 0$). But I is connected while S^0 is not. This contradicts the fact that the image of a connected space is connected.

As a corollary we get the following *fixed point theorem*.

10.2 Corollary

Suppose that $f: I \to I$ is a continuous function; then there exists some point $t \in I$ such that $f(t) = t$.

Proof If $f(0) = 0$ or $f(1) = 1$ we are finished. Suppose therefore that $f(0) > 0$ and $f(1) < 1$ and consider the function $g(t) = f(t) - t$. This is continuous and satisfies $g(0)\, g(1) < 0$. Thus by Lemma 10.1 we have $g(t) = 0$ for some $t \in I$ and hence $f(t) = t$ for some $t \in I$.

10.3 Corollary

Every continuous mapping of a circle to the real numbers sends at least one pair of diametrically opposite points to the same point.

Proof Suppose that $f(t) \neq f(-t)$ for all $t \in S^1$; then let h: $S^1 \rightarrow \mathbb{R}$ be the function $h(t) = f(t) - f(-t)$. Also, let e: $I \rightarrow S^1$ be given by $e(t) = \exp(\pi it)$. Clearly he is continuous. Now

$$he(0) = h(1) = f(1) - f(-1)$$
$$he(1) = h(-1) = f(-1) \, f(1) = -he(0).$$

So by Lemma 10.1 there is a point $t \in I$ such that $he(t) = 0$ and hence an $x \in S^1$ such that $h(x) = 0$ i.e. $f(x) = f(-x)$.

There is a physical interpretation of Corollary 10.3.

10.4   Corollary
At a given moment of time and a given great circle on the earth there is a pair of antipodal points with the same temperature.

Antipodal points are just diametrically opposite points. This result does generalize; see Chapter 20.

We come now to a precise statement of the first pancake problem.

10.5   Theorem
Let A and B be bounded subsets of the euclidean plane. Then there is a line in the plane which divides each region exactly in half by area.

Note that the two regions may overlap, i.e. the pancakes may overlap. Furthermore the regions need not be connected, i.e. the pancakes may be broken into several pieces.

Proof Let S be a circle with centre $(0,0) \in \mathbb{R}^2$ which contains both A and B (this is possible since A and B are bounded). By changing scales we may assume that S has diameter 1 unit. For each $x \in S$ consider the diameter D_x

Figure 10.1

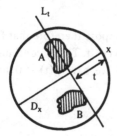

of S passing through x and let L_t be the line perpendicular to D_x passing through the point on D_x at a distance t from x ($t \in I$); see Figure 10.1.

Let $g_1(t)$ denote the area of that part of A which lies on the side of L_t nearest to x. Let $g_2(t)$ denote the area of the other part. (Note $g_1(0) = g_2(1) = 0$.) It is clear that g_1 and g_2 are continuous functions from I to $\mathbb{R}$. Define f: $I \to \mathbb{R}$ by

$$f(t) = g_2(t) - g_1(t).$$

It is continuous and satisfies $f(0) = -f(1)$, i.e. $f(0) f(1) \leq 0$. By Lemma 10.1 we know that there is some point $t \in I$ such that $f(t) = 0$. This point may not be unique. Because g_2 and $-g_1$ are monotone decreasing functions (this is obvious) so is $f = g_2 - g_1$. Thus $f(t) = 0$ on either a closed interval $[a,b]$ or at some unique point c. In the former case let $h_A(x) = \frac{1}{2}(a+b)$ while in the latter case $h_A(x) = c$. In other words a line perpendicular to D_x passing through the point distance $h_A(x)$ from x on D_x bisects the area of A. Note that

$$h_A(-x) = 1 - h_A(x).$$

Also note that $h_A: S^1 \to I$ is a continuous function (the usual trick: move x slightly and see what happens to $h_A(x)$).

In an identical fashion we define a function $h_B: S^1 \to I$ by using B instead of A. Now define h: $S^1 \to \mathbb{R}$ by

$$h(x) = h_A(x) - h_B(x)$$

which is continuous because h_A and h_B are. Now, we have $h(x) = -h(-x)$ for all $x \in S^1$. But also by Corollary 10.3 there is some point $y \in S^1$ such that $h(y) = h(-y)$. Thus $h(y) = 0$, $h_A(y) = h_B(y)$ and the line perpendicular to D_y passing through the point on D_y at a distance $h_A(y)$ from y bisects the area of A and the area of B.

The above theorem generalizes to higher dimensions, i.e. n bounded regions in $\mathbb{R}^n$; for n = 3 see Chapter 20.

The second pancake problem is now stated precisely.

10.6 Theorem

If A is a bounded region in the plane then there exists a pair of perpendicular lines which divide A into four parts each of the same area.

Proof As in the proof of Theorem 10.5 we enclose A within a circle S centre $(0,0) \in \mathbb{R}^2$ and diameter 1 unit. For each $x \in S$ let L_x be the line perpendicular to D_x which meets D_x at a distance $h_A(x)$ from x (in particular L_x bisects the area of A). Let y be the point on S at an angle $\frac{1}{2}\pi$ from x measured counterclockwise (i.e. $y = ix = x\sqrt{(-1)}$). Now let M_x be the line perpendicular to D_y which meets D_y at a distance $h_A(y)$ from y (again M_x bisects the area of A). Finally denote the four parts of A, working counter-clockwise, by $A_1(x)$, $A_2(x)$, $A_3(x)$, $A_4(x)$; see Figure 10.2. Note that if we denote area $(A_i(x))$ by $g_i(x)$ then

$$g_1(x) + g_2(x) = g_3(x) + g_4(x),$$
$$g_4(x) + g_1(x) = g_2(x) + g_3(x),$$

which gives $g_1(x) = g_3(x)$ and $g_2(x) = g_4(x)$. Of course, each of g_1, g_2, g_3 and g_4 is a continuous function from S^1 to $\mathbb{R}$. Let f be the continuous function given by

$$f(x) = g_1(x) - g_2(x) = g_3(x) - g_4(x)$$

Notice that

$$
\begin{aligned}
f(ix) &= g_1(ix) - g_2(ix) \\
&= g_2(x) - g_3(x) \\
&= g_2(x) - g_1(x) \\
&= -f(x).
\end{aligned}
$$

Now apply Lemma 10.1 to the function $f\sqrt{e}\colon I \to \mathbb{R}$, where $\sqrt{e}\colon I \to S^1$ is given by $\sqrt{e}(t) = \exp(\pi it/2)$, to obtain the required result.

Figure 10.2

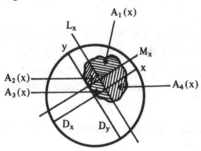

The solution to the pancake problems is an existence theorem; it asserts that a cut of the required kind exists but it does not tell us where to make the cut. In general the precise position of the cut may be difficult to find; we give, by way of an exercise, one example where it is easy to find.

10.7   Exercises

(a) Suppose that there are two pancakes on a plate. If one of them has the shape of a regular 2n-gon and the other the shape of a regular

2m-gon where would you make a cut with one stroke of a knife in order to divide both exactly in half?

(b) (An alternative proof of Theorem 10.5.) Using the notation of 10.5 first show that for $x \in S^1$ there is a line L_x perpendicular to D_x which divides A in half. This line L_x divides B into two parts; let $k_1(x)$, $k_2(x)$ be the areas of those parts of B that are respectively closest to, farthest from x. Let $k(x) = k_1(x) - k_2(x)$. Show that k: $S^1 \rightarrow \mathbb{R}$ is continuous and deduce Theorem 10.5.

Manifolds and surfaces

In this chapter we look at a special class of topological spaces: ones that locally look just like euclidean spaces.

11.1 Definition

Let n be a non-negative integer. An *n-dimensional manifold* is a Hausdorff space in which each point has an open neighbourhood homeomorphic to the open n-dimensional disc $\overset{\circ}{D}{}^n = \{\ x \in \mathbb{R}^n;\ \|x\| < 1\ \}$. Note that $\overset{\circ}{D}{}^n \cong \mathbb{R}^n$, so that we could equally require that each point has a neighbourhood homeomorphic to $\mathbb{R}^n$. For brevity we talk about an *n-manifold*.

Since $\mathbb{R}^0$ is just a single point it follows that any space X with the discrete topology is a 0-manifold. (A space with the discrete topology is Hausdorff and for $x \in X$ we can choose $\{\ x\ \}$ as the open set containing x which is homeomorphic to $\mathbb{R}^0$.) Apart from the 0-manifolds perhaps the simplest example of an n-manifold is $\mathbb{R}^n$ or $\overset{\circ}{D}{}^n$ itself. Also, any open subset of $\mathbb{R}^n$ is an n-manifold: if U is an open subset of $\mathbb{R}^n$ and $u \in U$ then there exists an $\epsilon > 0$ such that $u \in B_\epsilon(u) \subseteq U \subseteq \mathbb{R}^n$, and of course $B_\epsilon(u) \cong \overset{\circ}{D}{}^n$.

The circle S^1 is a 1-manifold. To see this let $S^1 \subseteq \mathbb{C}$ be given by

$$\{\ \exp(2\pi i t)\ ;\ t \in I\ \}\ .$$

If $x = \exp(2\pi i \theta) \in S^1$ then

$$
\begin{aligned}
x \in S^1 - \{\ -x\ \} &= S^1 - \{\ \exp(2\pi i(\theta - \tfrac{1}{2}))\ \} \\
&= \{\ \exp(2\pi i t)\ ;\ \theta - \tfrac{1}{2} < t < \theta + \tfrac{1}{2}\ \} \\
&\cong (\theta - \tfrac{1}{2},\ \theta + \tfrac{1}{2}) \\
&\cong (0,1) \cong \overset{\circ}{D}{}^1
\end{aligned}
$$

so that each point has a neighbourhood homeomorphic to $\overset{\circ}{D}{}^1$. Clearly S^1 is Hausdorff and so S^1 is a 1-manifold. More generally S^n is an n-manifold. To see this we introduce the idea of stereographic projection which is in fact a homeomorphism from $S^n - \{\ (0,0,...,0,1)\ \}$ to $\mathbb{R}^n$. We define it as follows: for $x \in S^n - \{\ (0,0,...,0,1)\ \}$, draw a straight line from $(0,0,...,0,1)$ to x in $\mathbb{R}^{n+1}$ extended to meet $\mathbb{R}^n = \{ (x_1, x_2, ..., x_n, x_{n+1}) \in \mathbb{R}^{n+1}; x_{n+1} = 0 \} \subseteq \mathbb{R}^{n+1}$. The point of intersection defines uniquely $\varphi(x)$. See Figure 11.1.

Figure 11.1

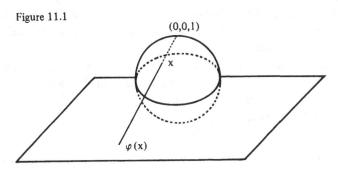

It is not difficult, intuitively at least, to see that φ is continuous and bijective. Also $\psi = \varphi^{-1}$ can be easily defined and seen to be continuous. A precise formula for φ is not difficult to obtain: just write down the equation for a straight line in $\mathbb{R}^{n+1}$ passing through $(0,0,...,0,1)$ and x, then find the point in this line for which $x_{n+1} = 0$. The reader will quickly find that

$$\varphi(x_1, x_2, ..., x_{n+1}) = \left(\frac{x_1}{1 - x_{n+1}}, \frac{x_2}{1 - x_{n+1}}, ..., \frac{x_n}{1 - x_{n+1}} \right)$$

An inverse $\psi : \mathbb{R}^n \to S^n - \{(0,0,...,0,1)\}$ is given by

$$\psi(x_1, x_2, ..., x_n) = \frac{1}{1 + \|x\|^2} (2x_1, 2x_2, ..., 2x_n, \|x\|^2 - 1).$$

We leave it for the reader to check that φ and ψ are continuous and that $\varphi\psi = 1$, $\psi\varphi = 1$.

It follows that any point $x \in S^n - \{ (0,0,...,0,1) \}$ has a neighbourhood, namely $S^n - \{ (0,0,...,0,1) \}$ itself, which is homeomorphic to $\overset{\circ}{D}{}^n$. Finally, the point $(0,0,...,0,1)$ has the neighbourhood $S^n - \{ (0,0,...,0,-1) \}$ which is homeomorphic to $\mathbb{R}^n$ via the map φ', where

$$\varphi'(x_1, x_2, ..., x_{n+1}) = \left(\frac{x_1}{1 + x_{n+1}}, \frac{x_2}{1 + x_{n+1}}, ..., \frac{x_n}{1 + x_{n+1}} \right)$$

It follows that S^n is indeed an n-manifold.

Another way to see that S^n is an n-manifold is to first look at the point $(0,0,...,0,1) \in S^n$ and the neighbourhood U of $(0,0,...,0,1)$ given by

$$U = \{ (x_1, x_2, ..., x_{n+1}) \in S^n; x_{n+1} > 0 \}.$$

This neighbourhood U is homeomorphic to $\overset{\circ}{D}{}^n$ by orthogonal projection, in other words by the map $U \to \overset{\circ}{D}{}^n \subseteq \mathbb{R}^n$ given by $(x_1, x_2, ..., x_{n+1}) \to$

$(x_1, x_2, ..., x_n)$. In general, for $x \in S^n$, we take U_x to be

$$U_x = \{ y \in S^n; \|x - y\| < \sqrt{2} \}$$

which is clearly an open neighbourhood of $x \in S^n$. Orthogonal projection onto the n-plane in $\mathbb{R}^{n+1}$ passing through O and orthogonal to the line through O and x produces a homeomorphism between U_x and $\overset{\circ}{D}^n$ and shows that S^n is an n-manifold.

The fact that an n-manifold, by definition, is Hausdorff is important. One might ask: If X is a space in which each point has a neighbourhood homeomorphic to $\mathbb{R}^n$ then is X Hausdorff? The answer is no, as a simple example will show. Let X be the set

$$X = \{ x \in R; -1 < x \leq 2 \}$$

with the topology $\mathcal{U}$ where $U \in \mathcal{U}$ if $U = \emptyset$, $U = X$ or U is an arbitrary union of sets of the form

$$(\alpha, \beta) \qquad\qquad -1 \leq \alpha < \beta \leq 2,$$
$$(\alpha, 0) \cup (\beta, 2] \qquad -1 \leq \alpha < 0, -1 \leq \beta < 2.$$

Note that X does not have the subspace topology induced by R because sets of the form $(\beta, 2]$ are not open in X. Intuitively the correct picture of X is given in Figure 11.2 (*a*) or (*b*). This is because $\{ 2 \}$ is arbitrarily close to $\{ 0 \}$ (i.e. any open set containing $\{ 2 \}$ contains $(\alpha, 0)$ for some α). Clearly X is not Hausdorff because any open neighbourhood of $\{ 2 \}$ intersects every open neighbourhood of $\{ 0 \}$. On the other hand every point in X has a neighbourhood homeomorphic to $\mathbb{R}^1$. If $x \in X$ and $x \neq 2$ then this is clear. If $x = 2$ then

$$N = (-\tfrac{1}{2}, 0) \cup (3/2, 2]$$

is a neighbourhood of $\{ 2 \}$ which is homeomorphic to $\overset{\circ}{D}^1$ by the map f: $N \to (-1,1) = \overset{\circ}{D}^1$ where

$$f(y) = \begin{cases} 2y & \text{if } -\tfrac{1}{2} < y < 0, \\ 4-2y & \text{if } 3/2 < y \leq 2. \end{cases}$$

The reader should check that f is continuous and bijective with an inverse $g: (-1,1) \to N$ given by

Figure 11.2

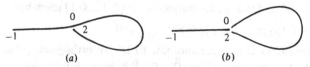

(*a*)   (*b*)

$$g(x) = \begin{cases} \tfrac{1}{2}x & \text{if } -1 < x < 0, \\ 2 - \tfrac{1}{2}x & \text{if } 0 \le x < 1. \end{cases}$$

Thus the Hausdorff condition in Definition 11.1 is not at all superfluous. It rules out spaces as illustrated in Figure 11.2, spaces that do not intuitively feel locally like euclidean spaces. Perhaps another reason for including the Hausdorff condition is that we tend to think of n-manifolds as subspaces of some euclidean space $\mathbb{R}^N$, N large (which locally are like $\mathbb{R}^n$). In this case the Hausdorff condition is inherited from the surrounding space $\mathbb{R}^N$. In fact there is a theorem which states that if M is a nice n-manifold (for example compact) then M is homeomorphic to a subspace of some euclidean space $\mathbb{R}^N$. See Exercises 11.2(f) and (g) for the case of a compact manifold.

For further examples of manifolds note that if M is an m-manifold and N is an n-manifold then the product $M \times N$ is an (m + n)-manifold because $\overset{\circ}{D}^m \times \overset{\circ}{D}^n \cong \mathbb{R}^m \times \mathbb{R}^n \cong \mathbb{R}^{m+n} \cong \overset{\circ}{D}^{m+n}$ and products of Hausdorff spaces are Hausdorff. Thus $S^1 \times S^1$ is a 2-manifold and more generally $\underbrace{S^1 \times S^1 \times \dots \times S^1}_{n}$ is an n-manifold.

The space $\mathbb{R}P^n$ is an n-manifold. To see this consider the map p: $S^n \to \mathbb{R}P^n$ sending $x \in S^n$ to the pair $\{x, -x\} \in \mathbb{R}P^n$. Let U_x be an open neighbourhood of $x \in S^n$ that is homeomorphic to $\overset{\circ}{D}^n$ and has diameter less than $\sqrt{2}$. In that case $p(U_x)$ is an open neighbourhood of $\{x, -x\} \in \mathbb{R}P^n$ homeomorphic to $\overset{\circ}{D}^n$. This is because p is a continuous open mapping (Theorem 5.12) and if U is a small enough region in S^n then $p|U : U \to p(U)$ is bijective. More generally let X be a G-space where G is a finite group. We say that G acts *freely* on X if $g \cdot x \neq x$ for all $x \in X$ and all $g \in G$, $g \neq 1$. If G acts freely on X then if X is a compact n-manifold so is X/G; conversely if X/G is an n-manifold then so is X. We leave details for the reader.

For another example consider the identification space M depicted in Figure 11.3, which consists of an octagonal region X with its edges identified as illustrated. Let p: $X \to M$ denote the natural projection map.

Figure 11.3

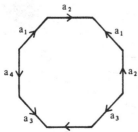

If $x \in M$ is such that $p^{-1}(x)$ is in the interior of X then clearly x has a neighbourhood homeomorphic to $\overset{\circ}{D}^2$; in fact $p(\overset{\circ}{X})$ is such a neighbourhood. If $x \in M$ is such that $p^{-1}(x)$ belongs to an edge of X but not a vertex of X then again it is not difficult to see that x has a neighbourhood homeomorphic to $\overset{\circ}{D}^2$; see Figure 11.4.

Figure 11.4

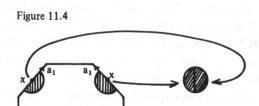

Finally if $p^{-1}(x)$ is a vertex of X then a neighbourhood N_x of x which is homeomorphic to $\overset{\circ}{D}^2$ is depicted in Figure 11.5 ($p^{-1}(N_x)$ consists of points in X of distance less than ϵ from $p^{-1}(x)$ for some suitable $\epsilon > 0$).

Intuitively it is not difficult to see that M is Hausdorff and the reader should have no difficulty seeing this. For the algebraic minded we give the following proof. Let A denote the 'edges' of X. Write A as $\overset{8}{\underset{i=1}{\cup}} A_i \cup Y$ where the A_i are the (closed) edges of X, and Y denotes the eight vertices of X. Let C be a closed subset of X, then

$$p^{-1}p(C) = p^{-1}p((C - A) \cup (C \cap Y) \cup \overset{8}{\underset{i=1}{\cup}} (C \cap A_i))$$

$$= (C - A) \cup p^{-1}p(C \cap Y) \cup \overset{8}{\underset{i=1}{\cup}} p^{-1}p(C \cap A_i)$$

$$= (C - A) \cup \epsilon Y \cup \overset{8}{\underset{i=1}{\cup}} ((C \cap A_i) \cup B_i)$$

Figure 11.5

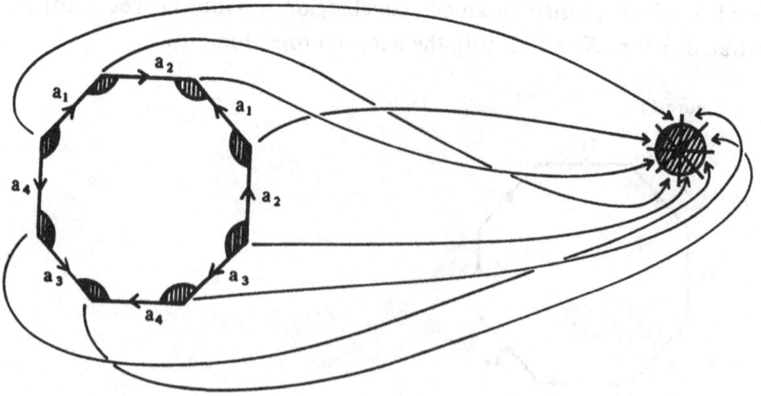

where $\epsilon Y = Y$ if $C \cap Y$ is non-empty and $\epsilon Y = \emptyset$ if $C \cap Y$ is empty. The set B_i is a subspace of A homeomorphic to $C \cap A_i$. In fact, if the edge A_i is identified with the edge A_j in M then B_i is the subspace of A_j homeomorphic to $C \cap A_i$ for which $p(B_i) = p(C \cap A_i)$. (Note that $p^{-1}p(C \cap A_i) \cap A_j = B_i \cup (\epsilon Y \cap A_j)$.) Thus we see that

$$p^{-1}p(C) = C \cup \epsilon Y \cup \bigcup_{i=1}^{8} B_i$$

See Figure 11.6.

Figure 11.6

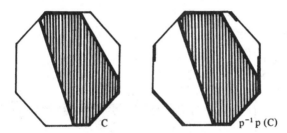

$$C \qquad\qquad p^{-1}p(C)$$

The subspaces B_i, $i = 1, 2, ..., 8$, are closed subsets of X because B_i is homeomorphic to $C \cap A_i$ which is a closed set, and so B_i is a closed subset of A_j (some j) and hence a closed subset of X. It follows that $p^{-1}p(C)$ is a closed set and hence so is $p(C)$ by the definition of a quotient topology. We deduce that p: $X \to M$ is a closed mapping. Since X is obviously a compact Hausdorff space it follows by Theorem 8.11 that M is also a compact Hausdorff space and so M is indeed a 2-manifold.

We can actually perform the identifications of Figure 11.3 within our three-dimensional world. This is depicted in Figure 11.7. The end result is called a *double torus*.

Another way of realizing the 2-manifold M of Figure 11.3 is to first remove an open disc neighbourhood of a point as indicated in Figure 11.8(a). Now identify the two edges denoted by a_1 to give (c). Consider the shaded region Y in (d). This is homeomorphic to the subspace of $\mathbb{R}^2$ illustrated in (e). The function f,

$$f(x,y) = \begin{cases} (x(a + 2y(b - a))/b, y) & \text{if } 0 \le x \le b \text{ and } y \le 0, \\ \left(\dfrac{(x - b)(1 - a - 2y(b - a))}{1 - b} + a + 2y(b - a), y\right) \\ \qquad\qquad\qquad\qquad\qquad \text{if } b \le x \le 1 \text{ and } y \le 0, \\ (xa/b, y) & \text{if } y \ge 0 \end{cases}$$

(where $0 < a \leq b < 1$) is a homeomorphism between the spaces illustrated in Figure 11.8(e) and (f). Notice that f is the identity on the three edges of these regions not containing the bump. Thus we get a homeomorphism from Y to Y which is the identity on the three edges of Y not containing a_2. Using this homeomorphism on Y and the identity on the non-shaded part of (d) we get a homeomorphism between (d) and (g). Thus (c) and (h) are homeomorphic spaces. In a similar way (h) and (k) are homeomorphic, using the shaded regions of (i) and (j) to define a homeomorphism. Making the identification on a_2 gives (l). In a similar way we obtain (m) which is homeomorphic to the original (a).

Figure 11.7

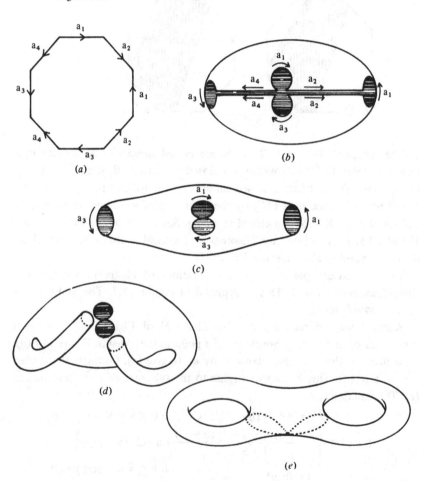

Figure 11.8

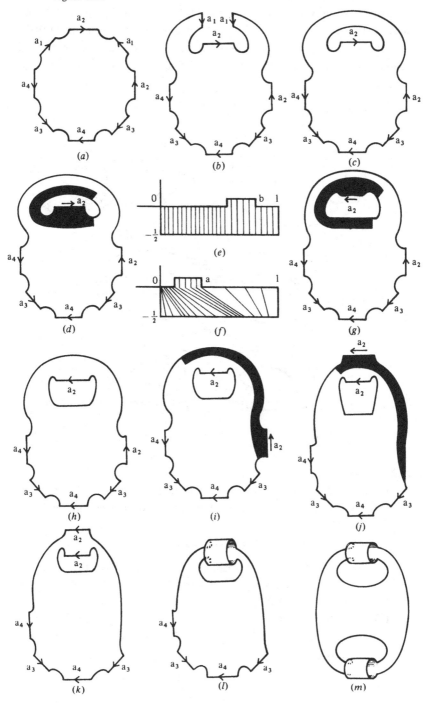

To continue we go over to Figure 11.9 and after some simple stretching homeomorphisms we arrive at Figure 11.9(c). Finally, replacing the disc neighbourhood that we removed gives us the double torus in Figure 11.9(d).

It turns out that all compact 2-manifolds can be obtained as the identification space of some polygonal region; we shall return to this later on in this chapter.

Figure 11.9

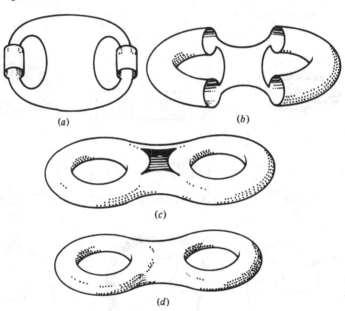

(a)   (b)

(c)

(d)

11.2   Exercises

(a) Show that an open subset of an n-manifold is also an n-manifold.

(b) Let $CP^n = S^{2n+1} / \sim$ where $\sim$ is the equivalence relation on $S^{2n+1} \subseteq C^{n+1}$ given by

$$x \sim y \Leftrightarrow x = \exp(2\pi it)\, y \text{ for some } t \in I.$$

Prove that CP^n is a 2n-manifold. (Note that $\sim$ identifies circles in S^{2n+1} to a single point; e.g. $\{\,(\exp(2\pi it), 0,0,...,0);\ t \in I\,\}$ represents a point in CP^n.)

(c) Let p be a positive integer and let $L_p = S^{2n+1} / \sim$ where $\sim$ is the equivalence relation on $S^{2n+1} \subseteq C^{n+1}$ given by

$$x \sim y \Leftrightarrow x = \exp(2\pi in/p)\, y,\ n = 0,1,...,p-1.$$

Prove that L_p is a $(2n+1)$-manifold. $(L_p = S^{2n+1}/\mathbb{Z}_p$ where $\mathbb{Z}_p$ acts freely on S^{2n+1} in an obvious way.)

(d) Let X be a G-space where G is a finite group that acts freely on X. Prove that if X is a compact n-manifold then so is X/G. Prove also that if X/G is a manifold then so is X.

(e) Prove that if M is an n-manifold then each point of M has a neighbourhood homeomorphic to the closed n-disc D^n.

(f) Suppose that M is a compact n-manifold. Prove that M is homeomorphic to a subspace of some euclidean space $\mathbb{R}^N$. (Hint: Since M is compact there is a finite cover $\{ D_1, D_2, ..., D_m \}$ of M and homeomorphisms $h_i: D_i \to \overset{\circ}{D}{}^n$. Use Exercises 8.14(j) and (k) to get homeomorphisms $M/(M-D_i) \cong (D_i)^\infty \cong (\overset{\circ}{D}{}^n)^\infty \cong S^n$. Since M is compact and Hausdorff and $M-D_i$ is closed the projection $p_i: M \to M/(M-D_i)$ is continuous so that we get continuous maps $f_i: M \to S^n$. Define $f: M \to (S^n)^m$ by $f(x) = (f_1(x), f_2(x), ..., f_m(x))$. Finally $(S^n)^m \subseteq (\mathbb{R}^n)^m = \mathbb{R}^{nm}$.)

(g) Let M be an n-manifold and let D be a subspace of M which is homeomorphic to $\overset{\circ}{D}{}^n$. Since $\overset{\circ}{D}{}^n \cong \mathbb{R}^n \cong S^n - \{ (0,0,...,0,1) \}$ we have a homeomorphism

$g: D \to S^n - \{ (0,0,...,0,1) \}$.

Define $f: M \to S^n$ by

$$f(x) = \begin{cases} g(x) & \text{if } x \in D, \\ (0,0,...,0,1) & \text{if } x \in M-D. \end{cases}$$

Prove that f is continuous. Use this result to reprove the result in (f).

Our prime interest is in compact connected manifolds. All compact connected 0-manifolds are homeomorphic to each other. The circle S^1 is a compact connected 1-manifold. In fact, up to homeomorphism, S^1 is the only compact connected 1-manifold. The proof of this is not very difficult and we shall give an outline of it. The first step (perhaps the hardest, but intuitively plausible) is to use compactness to show that if M is a compact connected 1-manifold then M can be subdivided in a 'nice' way into a finite number of regions each homeomorphic to the unit interval I. If we call homeomorphic images of I *arcs* and the image of $\{ 0,1 \}$ the *vertices* of the arcs then by 'nice' we mean that no arc intersects itself, and whenever two arcs intersect then they do so at one or two vertices. (The idea is to (i) cover M by neighbourhoods of points homeomorphic to $\overset{\circ}{D}{}^1 \cong \overset{\circ}{I}$, (ii) choose finitely many of these by compactness, (iii) choose smaller neighbourhoods homeomorphic to I which still cover M, and finally, (iv) use the definition of a

1-manifold to show that M has a 'nice' subdivision.) Clearly in such a nice subdivision of M into arcs and vertices each vertex is a vertex of exactly two distinct arcs and each arc has two distinct vertices. (If a vertex is the vertex of one or of more than two arcs then that vertex does not have a neighbourhood homeomorphic to $\overset{\circ}{D}^1$.) Suppose that M has more than two arcs. Let A_1, A_2 be a pair of arcs in M that meet at a vertex a.

$$\bullet\!\!\!\underset{\displaystyle a}{\overset{\displaystyle A_1}{\rule{4cm}{0.4pt}}}\!\!\!\bullet\!\!\!\overset{\displaystyle A_2}{\rule{4cm}{0.4pt}}\!\!\!\bullet$$

Let $h_1: A_1 \to I$, $h_2: A_2 \to I$ be homeomorphisms defining A_1 and A_2 as arcs. We may suppose that $h_1(a) = 1$ and $h_2(a) = 0$; otherwise compose h_1 and/or h_2 with the homeomorphism f: $I \to I$ given by f(t) = 1-t. Define g: $A_1 \cup A_2 \to I$ by

$$g(x) = \begin{cases} \tfrac{1}{2} h_1(x) & \text{if } x \in A_1, \\[2mm] \tfrac{1}{2} + \tfrac{1}{2} h_2(x) & \text{if } x \in A_2. \end{cases}$$

This is well defined and easily seen to be bijective. To see that g is continuous note first that A_1 and A_2 are closed subsets of $A_1 \cup A_2$ and of M. Let C be a closed subset of I, then

$$g^{-1}(C) = h_1^{-1}([0, \tfrac{1}{2}] \cap C) \cup h_2^{-1}([\tfrac{1}{2}, 1] \cap C)$$

which is clearly closed in $A_1 \cup A_2$, and so g is indeed continuous. That g is a

Figure 11.10

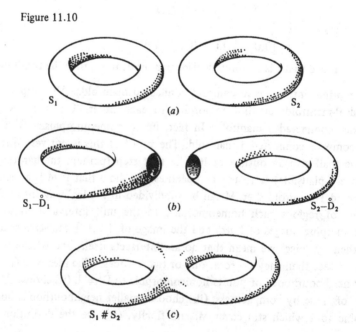

homeomorphism follows easily. We can therefore replace arcs A_1, A_2 by one arc A. We now have a subdivision of M with one fewer arc and one fewer vertex. Continuing in this way we end up with a subdivision of M consisting of two arcs and hence with two vertices. Thus M is homeomorphic to two copies of I with the ends glued together pairwise. The result is that M is homeomorphic to the circle S^1.

Compact connected 2-manifolds are called *surfaces*. Examples of surfaces are the sphere S^2, the torus $T = S^1 \times S^1$, the real projective plane $\mathbb{R}P^2$ and the double torus described earlier on in this chapter. The first three examples are basic in the sense that any surface can be obtained from these three surfaces by a process called 'connected sum'. Let S_1 and S_2 be two disjoint surfaces, their *connected sum* $S_1 \# S_2$ is formed by removing small open discs from each surface and glueing along the boundaries of the resulting holes. See Figure 11.10.

For a more rigorous definition first choose $D_1 \subseteq S_1$ and $D_2 \subseteq S_2$ so that D_1 and D_2 are homeomorphic to D^2. That such regions exist is easy to see: Let x be some point of a surface, then x has a neighbourhood N and a homeomorphism h: $N \to \overset{\circ}{D}^2$. The subspace $h^{-1}(D^2{}_{1/2}) \subseteq N$, where $D^2{}_{1/2} \subseteq \overset{\circ}{D}^2$ is a closed disc of radius ½, is homeomorphic to D^2 by the homeomorphism

$$h^{-1}(D^2{}_{1/2}) \to D^2,$$
$$y \to 2h(y).$$

Returning to the definition of $S_1 \# S_2$ let $D_1 \subseteq S_1$ and $D_2 \subseteq S_2$ be subspaces homeomorphic to D^2 and let $h_1: D_1 \to D^2$, $h_2: D_2 \to D^2$ be homeomorphisms. Define $S_1 \# S_2$ by

$$\frac{(S_1 - \overset{\circ}{D}_1) \cup (S_2 - \overset{\circ}{D}_2)}{\sim}$$

where $\sim$ is an equivalence relation which is non-trivial only on $\partial(S_1 - \overset{\circ}{D}_1) \cup \partial(S_2 - \overset{\circ}{D}_2) = \partial D_1 \cup \partial D_2$; there it is given by $x \sim h_2^{-1}h_1(x)$ for $x \in \partial D_1$. It is possible to show that the definition of connected sum is independent of the various choices of discs D_1, D_2 and homeomorphisms h_1, h_2. It is not difficult to see that $S_1 \# S_2$ is a surface: the only neighbourhoods of points that we need to look at are those in ∂D_1 or ∂D_2. Details are left for the reader.

The double torus is the connected sum of two tori; see Figure 11.10. The Klein bottle is the connected sum of two projective planes. This can be seen quite easily from Chapter 5 but a geometric proof is illustrated in Figure 11.11. We start with two projective planes as in Figure 11.11(a); then we remove two open discs as indicated in (b). The result is homeomorphic to the space in (c). Glueing together (i.e. taking connected sum) gives (d). Finally we make a cut as in (e) to give us the identification space (f); rearranging

Figure 11.11

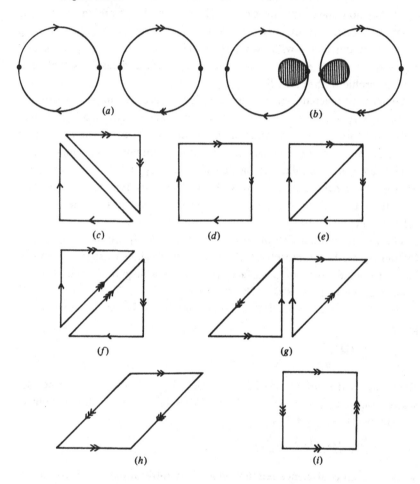

gives (*g*), identifying gives (*h*) and a simple homeomorphism gives (*i*) which is a Klein bottle.

The fact that all surfaces can be obtained from the sphere S^2, the torus $T = S^1 \times S^1$ and the real projective plane $\mathbb{R}P^2$ via connected sums gives the so-called *classification theorem of surfaces.*

11.3   Theorem

Let S be a surface. S is homeomorphic to precisely one of the following surfaces:

$$S^2 \,\#\, \underbrace{T \,\#\, T \,\#\, ... \,\#\, T}_{m} \qquad\qquad (m \geq 0),$$

$$S^2 \,\#\, \underbrace{\mathbb{R}P^2 \,\#\, \mathbb{R}P^2 \,\#\, ... \,\#\, \mathbb{R}P^2}_{n} \qquad (n \geq 1).$$

The proof breaks down into two parts. The first part is to show that any surface is homeomorphic to at least one of the surfaces listed in Theorem 11.3. We shall not give full details of this part, but merely a brief outline later on in this chapter. The second part of the proof is to show that no two of the surfaces listed in the theorem are homeomorphic; this we do rigorously in Chapter 26.

Taking the connected sum with a torus is often referred to as *sewing on a handle* (where a *handle* is just a torus with an open disc removed); the reason should be obvious. (See Figure 11.12(*e*), (*f*) for an example.) Sometimes we talk about *sewing on a cylinder*: to do this remove two open discs in the surface and sew on the cylinder as indicated in Figure 11.12(*a*); it is important to do this sewing on correctly. Sewing on a cylinder incorrectly (i.e. reversing the arrow on one of the boundary circles) is equivalent to taking the connected sum with a Klein bottle K.

Figure 11.12

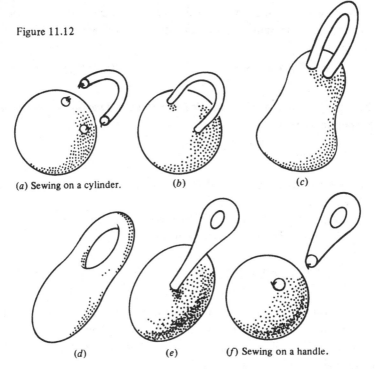

(*a*) Sewing on a cylinder.   (*b*)   (*c*)

(*d*)   (*e*)   (*f*) Sewing on a handle.

Taking the connected sum with the real projective plane is often referred to as *sewing on a Möbius strip* (see Figure 11.13). This is simply because a projective plane with an open disc removed is just a Möbius strip (see Chapter 5).

Figure 11.13. Sewing on a Mobius strip.

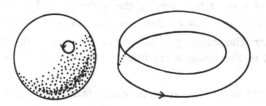

The surfaces obtained after taking a connected sum with $\mathbb{R}P^2$ are special in they are (commonly called) one-sided. This is because they contain a Möbius strip which as we saw in Chapter 5 has some strange properties. We say that a surface is *orientable* if it does not contain a Möbius strip within it. On the other hand we say that a surface is *non-orientable* if it does contain a Möbius strip within it.

Thus the Klein bottle and the real projective plane are both non-orientable surfaces while the sphere, torus and double torus are orientable surfaces. The surface

$$S^2 \underbrace{\# T \# T \# ... \# T}_{m} \qquad (m \geq 0),$$

which we abbreviate to $S^2 \# mT$, is said to be the *standard orientable surface of genus m*. The surface

$$S^2 \# \underbrace{\mathbb{R}P^2 \# \mathbb{R}P^2 \# ... \# \mathbb{R}P^2}_{n} \qquad (n \geq 1),$$

which we abbreviate to $S^2 \# n\mathbb{R}P^2$, is said to be the *standard non-orientable surface of genus n*.

A natural question to ask is: What surface do we get if we take connected sums of tori and projective planes? In other words, to which standard surface is

$$\underbrace{T \# T \# ... \# T}_{m} \# \underbrace{\mathbb{R}P^2 \# \mathbb{R}P^2 \# ... \# \mathbb{R}P^2}_{n} = mT \# n\mathbb{R}P^2$$

homeomorphic?, where m, n ≥ 1. Such a surface is clearly non-orientable and so, if we assume Theorem 11.3 then $mT \# n\mathbb{R}P^2$ is homeomorphic to $k\mathbb{R}P^2$ for some k. We shall determine the value of k in the case m = n = 1 and leave the general case as an (easy) exercise.

11.4 Lemma

$$T \# \mathbb{R}P^2 \cong \mathbb{R}P^2 \# \mathbb{R}P^2 \# \mathbb{R}P^2.$$

Proof Denote $T \# \mathbb{R}P^2$ by S_1 and $\mathbb{R}P^2 \# \mathbb{R}P^2 \# \mathbb{R}P^2$ by S_2. We shall first represent S_1 and S_2 as identification spaces. S_1 is the quotient space of a hexagonal region X; see Figure 11.14.

Note that all vertices of X are identified to one point in S_1 and there is a neighbourhood $\overset{\circ}{D}_1$ of this point in S_1 which is homeomorphic to $\overset{\circ}{D}^2$; see Figure 11.14(e). Removing this neighbourhood (Figure 11.15(a)) and making the identifications gives the space of Figure 11.15(c) which after a homeomorphism gives Figure 11.15(d). We shall describe a sequence of

Figure 11.14

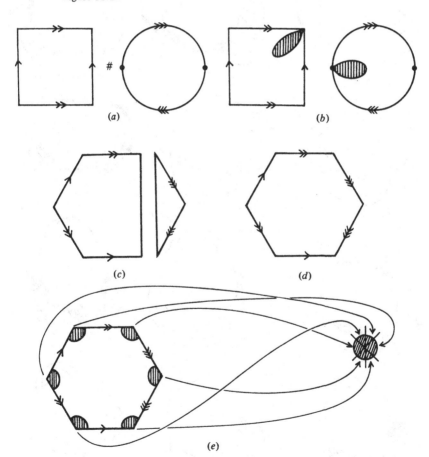

(a)

(b)

(c)

(d)

(e)

Figure 11.15

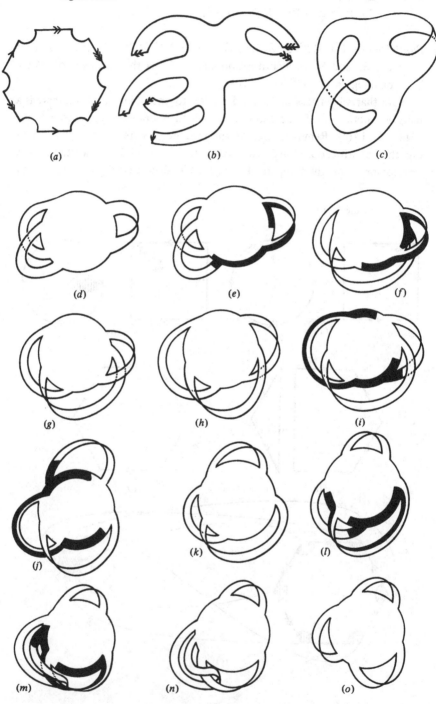

(a)   (b)   (c)

(d)   (e)   (f)

(g)   (h)   (i)

(j)   (k)   (l)

(m)   (n)   (o)

Figure 11.16

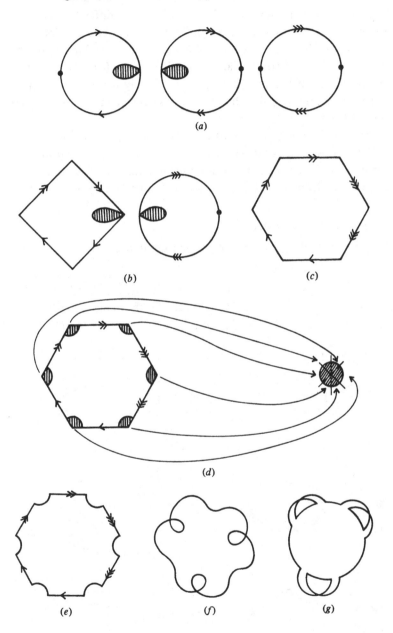

(a)

(b)   (c)

(d)

(e)   (f)   (g)

homeomorphisms that transform 11.15(d) to 11.15(o). To get from (d) to (g) consider the shaded region indicated in (e) and (f). By using the ideas described earlier in connection with Figure 11.8 it is not difficult to describe a homeomorphism between (d) and (g). That (g) and (h) are homeomorphic is easy to see. To show that (h) and (k) are homeomorphic we use the shaded regions of (i) and (j). Similarly to see that (k) and (n) are homeomorphic we use the shaded regions of (l) and (m). Finally it is easy to see that (n) and (o) are homeomorphic. Thus $S_1 - \mathring{D}_1$ is homeomorphic to the space illustrated in Figure 11.15(o).

On the other hand S_2 has the identification space representation given in Figure 11.16(c). Removing the neighbourhood $\mathring{D}_2$ (indicated in (d)) which is homeomorphic to an open disc and making the necessary identifications gives us the space of Figure 11.16(g). It is immediate that there is a homeomorphism

$$h: S_1 - \mathring{D}_1 \cong S_2 - \mathring{D}_2.$$

Furthermore it is clear that h induces a homeomorphism

$$\partial(S_1 - \mathring{D}_1) \cong \partial(S_2 - \mathring{D}_2).$$

This homeomorphism on the boundaries can be extended to a homeomorphism between D_1 and D_2: If h: $\partial(D_1) \to \partial(D_2)$ is the homeomorphism and h_1: $D_1 \cong D^2$, h_2: $D_2 \cong D^2$ then write $x \in D^2$ in polar coordinates as

Figure 11.17

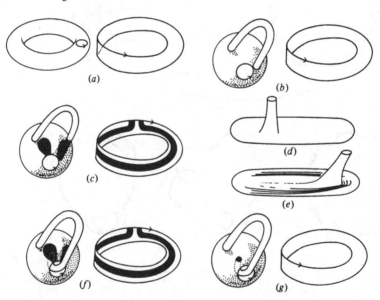

(a)

(b)

(c)

(d)

(e)

(f)

(g)

x = (r,t) where $0 \le r \le 1$ and $t \in \partial(D^2) = S^1$. Define H: $D_1 \to D_2$ by H(y) = $h_2^{-1}(r,h_2 hh_1^{-1}(t))$ where $h_1(y) = (r,t) \in D^2$. It is clear that $H|\partial D_1 = h$ and that H is a homeomorphism. Thus

$$S_1 = (S_1 - \overset{o}{D}_1) \cup D_1 \cong (S_2 - \overset{o}{D}_2) \cup D_2 = S_2$$

which completes the proof of Lemma 11.4.

There is another way of visualizing the homeomorphism between S_1 and S_2. We start by representing the connected sum $T \# \mathbb{R} P^2$ as a handle (a torus with a hole in it) together with a Möbius strip that has to be glued to it. This is illustrated in Figure 11.17(a) and it is clearly homeomorphic to the space in Figure 11.17(b). By performing the homeomorphism illustrated in (c) – (f) we arrive at Figure 11.17(g).

Consider a Klein bottle with a disc removed, Figure 11.18 shows what this space looks like.

It follows easily now that $S_1 \cong K \# \mathbb{R} P^2$ where K denotes the Klein bottle. But $K \cong \mathbb{R} P^2 \# \mathbb{R} P^2$ so that $S_1 \cong S_2$ as was to be shown.

Figure 11.18

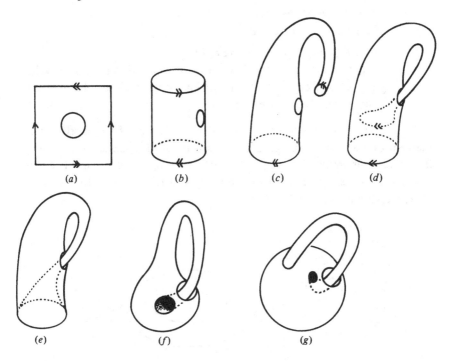

(a) (b) (c) (d)

(e) (f) (g)

11.5   Exercises

(a) Suppose S_1, S_2 and S_3 are surfaces. Show that

$S_1 \# S_2 \cong S_2 \# S_1$,

$(S_1 \# S_2) \# S_3 \cong S_1 \# (S_2 \# S_3)$,

$S^2 \# S_1 \cong S_1$.

Does the set of homeomorphism classes of surfaces form a group using connected sum as a law of composition? Why not?

(b) Let M_1, M_2 be disjoint, connected n-manifolds. Let D_1 and D_2 be subsets of M_1, M_2 respectively homeomorphic to D^n via h_1, h_2 say. Define the connected sum $M_1 \# M_2$ of M_1 and M_2 to be the quotient space

$$\frac{(M_1 - \mathring{D}_1) \cup (M_2 - \mathring{D}_2)}{\sim}$$

where $\sim$ identifies $x \in \partial(M_1 - \mathring{D}_1)$ with $h_2^{-1}h_1(x)$. Prove that $M_1 \# M_2$ is an n-manifold.

(c) Let $S = nT \# m\,RP^2$ with m, n ≥ 1. To which standard surface is S homeomorphic?

(d) Let S be a surface. Prove that S is homeomorphic to precisely one of the following surfaces:

$S^2 \# nT$, $RP^2 \# nT$, $K \# nT$

where K is the Klein bottle and $n \geq 0$.

(e) Suppose that the surface S is a G-space where $G = \mathbb{Z}_{2n+1}$, a cyclic group of odd order. Prove that S/G is a surface. Notice that it is not assumed that G acts freely on S.

We now give an outline of the first step in a proof of the classification theorem of surfaces. A subspace of a space is called a *simple closed curve* if it is homeomorphic to the circle S^1. If C is a simple closed curve in a surface S then we say that C *separates* S if S - C is not connected i.e. cutting along C disconnects S (see Figure 11.19).

Figure 11.19

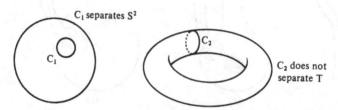

Let S be a surface which contains a simple closed curve C that does not separate S. It is possible to prove that a neighbourhood of C is either (i) a cylinder or (ii) a Möbius strip (Figure 11.20). Intuitively this should be clear.

Figure 11.20

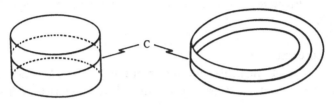

Now remove (the interior of) this cylinder or Möbius strip. In the first case sew on two discs to the two holes created, in the second case sew on one disc to the hole created. We thus get a surface M_1. Clearly M is just M_1 with a cylinder sewn on (correctly or incorrectly) or else with a Möbius strip sewn on. In other words

$$M = M_1 \# T, \ M = M_1 \# K \text{ or } M = M_1 \# \mathbb{R}P^2.$$

We now look at M_1 and find a simple closed curve in M_1 that does not separate M_1 (if one exists) and continue in the manner just described to obtain M_2 with $M_1 = M_2 \# T$, $M_1 = M_2 \# K$ or $M_1 = M_2 \# \mathbb{R}P^2$. Continuing in this manner we obtain after i steps M_i with

$$M = M_1 \# i_1 T \# i_2 K \# i_3 \ \mathbb{R}P^2$$

where $i_1 + i_2 + i_3 = i$. It turns out that after a finite number of steps (say $k \geq 0$) this process stops, i.e. every simple closed curve in M_k separates M_k. Finally we use a theorem which says that if M_k is a surface in which every simple closed curve in M_k separates M_k then M_k is homeomorphic to the sphere S^2.

Modulo the unproved assertions we see that M is homeomorphic to $S^2 \# \ell T \# mK \# n \ \mathbb{R}P^2$ for some $\ell, m, n \geq 0$ ($\ell + m + n = k$). By an easy application of Lemma 11.4 we see that $S^2 \# \ell T \# mK \# n \ \mathbb{R}P^2$ is homeomorphic to

$$S^2 \# \ell T \qquad \qquad \text{if } m + n = 0$$
$$S^2 \# (2\ell + 2m + n) \ \mathbb{R}P^2 \qquad \text{if } m + n > 0.$$

To complete the proof of the classification theorem we need to show that no two of the surfaces listed in Theorem 11.3 are homeomorphic; this we do in Chapter 26.

11.6 Exercises

(a) Show that a torus T has two distinct (but not disjoint) simple closed curves C_1, C_2 such that $T - (C_1 \cup C_2)$ is connected.

(b) Show that a torus T does *not* have three distinct simple closed
curves C_1, C_2, C_3 such that $T - (C_1 \cup C_2 \cup C_3)$ is connected.
(c) Generalize (a) and (b) to other surfaces.

We end this chapter with a result that we have already mentioned earlier
on: that a surface may be described as a quotient space of some polygonal
region in $\mathbb{R}^2$.

11.7   Theorem

If M is an orientable surface of genus $m \geq 1$ then M is the quotient
space of a 4m-sided polygonal region with identifications as indicated in
Figure 11.21(a).

If M is a non-orientable surface of genus $n \geq 1$ then M is the quotient
space of a 2n-sided polygonal region with identifications as indicated in
Figure 11.21(b).

Figure 11.21

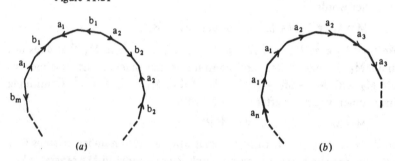

(a)   (b)

To prove this result we just need to show that mT and $n\mathbb{R}P^2$ take the
form indicated. We merely illustrate the cases $m \leq 2$, $n \leq 3$. T # T takes the
form as indicated in Figure 11.22. It should be clear how to proceed and
obtain the result for the orientable case. For the non-orientable case we have
Figure 11.23 for $\mathbb{R}P^2$ # $\mathbb{R}P^2$ and Figure 11.24 for $\mathbb{R}P^2$ # $\mathbb{R}P^2$ # $\mathbb{R}P^2$.
Again how to proceed should be clear.

Figure 11.22

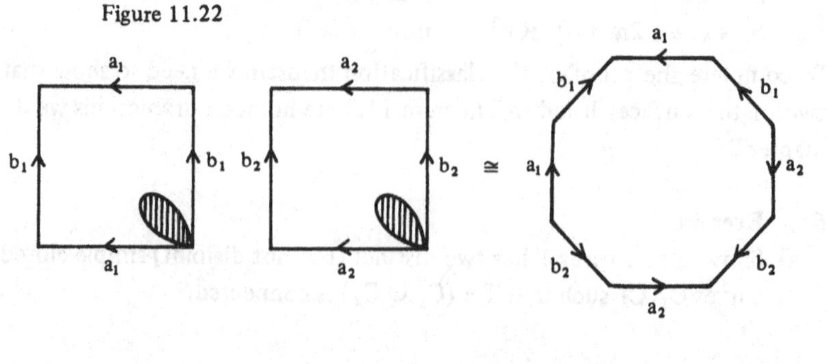

Figure 11.23

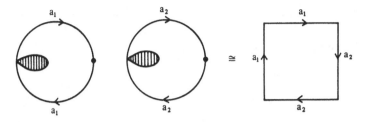

Figure 11.24

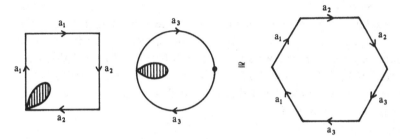

11.8   Exercises

(a) An *n-manifold-with-boundary* M is a Hausdorff space in which each point has an open neighbourhood homeomorphic to either $\mathbb{R}^n$ or the upper half-space of $\mathbb{R}^n$ i.e. $\{ (x_1,...,x_n) \in \mathbb{R}^n; x_n \geq 0 \}$. The set of points in M which have neighbourhoods homeomorphic to the upper half-space but which have no neighbourhoods homeomorphic to $\mathbb{R}^n$ is called the *boundary* of M. Prove that the boundary of an n-manifold-with-boundary is an n-1 manifold.

(b) A *surface-with-boundary* is a compact connected 2-manifold-with-boundary. Prove that the boundary of a surface-with-boundary is a disjoint union of a finite number of circles. Deduce that by glueing on a finite number of discs we may obtain from each surface-with-boundary a surface.

(c) A surface-with-boundary is said to be *orientable* if and only if it does not contain a Möbius strip. Prove that a surface-with-boundary is orientable if and only if the associated surface ((b) above) is orientable.

Paths and path connected spaces

In Chapter 9 we studied connectedness; in this chapter we study a somewhat similar and yet different concept of connectedness: path-connectedness. Before defining this we look at 'paths'. A continuous mapping $f: [0,1] \to X$ is called a *path* in X. The point $f(0)$ is called the *initial point* and $f(1)$ is called the *final* or *terminal point*. We say that f joins $f(0)$ and $f(1)$, and that f is a path from $f(0)$ to $f(1)$. (Some books use the term *arc* instead of path.)

Note that it is the mapping f that is the path and not the image $f([0,1])$ which is called a *curve* in X.

We usually think of $t \in [0,1]$ as time so that $f(t)$ is our position at time t.

The simplest example of a path is the constant path $\epsilon_x: [0,1] \to X$, defined by $\epsilon_x(t) = x$ for all $t \in [0,1]$, where x is some point of X. In this path we spend all our time at the same point $x \in X$.

There are two simple, but important, ways of obtaining new paths from old. These are given in the next lemma. The first associates to a path f a path $\bar{f}$ which essentially goes backwards along f. The second joins up two paths f, g (if possible) to give another path $f*g$.

12.1 Lemma
(a) If f is a path in X and $\bar{f}$ is defined by $\bar{f}(t) = f(1-t)$ then $\bar{f}$ is also a path in X.

(b) If f and g are two paths in X for which the final point of f coincides with the initial point of g then the function $f*g: [0,1] \to X$ defined by

$$(f*g)(t) = \begin{cases} f(2t) & \text{if } 0 \leq t \leq \tfrac{1}{2}, \\ g(2t-1) & \text{if } \tfrac{1}{2} \leq t \leq 1 \end{cases}$$

is a path in X.

Proof Part (a) is obvious while part (b) follows from the next result, the so-called *glueing lemma*.

12.2 Lemma

Let W, X be topological spaces and suppose that $W = A \cup B$ with A,B both closed subsets of W. If f: $A \rightarrow X$ and g: $B \rightarrow X$ are continuous functions such that $f(w) = g(w)$ for all $w \in A \cap B$ then h: $W \rightarrow X$ defined by

$$h(w) = \begin{cases} f(w) & \text{if } w \in A, \\ g(w) & \text{if } w \in B \end{cases}$$

is a continuous function.

Proof Note that h is well defined. Suppose that C is a closed subset of X, then

$$\begin{aligned} h^{-1}(C) &= h^{-1}(C) \cap (A \cup B) \\ &= (h^{-1}(C) \cap A) \cup (h^{-1}(C) \cap B) \\ &= f^{-1}(C) \cup g^{-1}(C). \end{aligned}$$

Since f is continuous, $f^{-1}(C)$ is closed in A and hence in W since A is closed in W. Similarly $g^{-1}(C)$ is closed in W. Hence $h^{-1}(C)$ is closed in W and h is continuous.

12.3 Definition

A space X is said to be *path connected* if given any two points x_0, x_1 in X there is a path in X from x_0 to x_1.

Note that by Lemma 12.1 it is sufficient to fix $x_0 \in X$ and then require that for all $x \in X$ there is a path in X from x_0 to x. (Some books use the term *arcwise connected* instead of path connected.)

For example $\mathbb{R}^n$ with the usual topology is path connected. The reason is that given any pair of points a, $b \in \mathbb{R}^n$ the mapping f: $[0,1] \rightarrow \mathbb{R}^n$ defined by $f(t) = tb + (1-t)a$ is a path from a to b. More generally any convex subset of $\mathbb{R}^n$ is path connected. A subset E of $\mathbb{R}^n$ is *convex* if whenever a, $b \in E$ then the set $\{ tb + (1-t) a; 0 \leq t \leq 1 \}$ is contained in E, i.e. E is convex if the straight-line segment joining any pair of points in E is in E itself. See Figure 12.1 for an example of a convex and of a non-convex subset of $\mathbb{R}^2$.

Figure 12.1

A convex subset.

A non-convex subset.

In particular any interval in R^1 is path connected.

The next few results 12.4 – 12.7 are analogous to the results 9.4 – 9.7.

12.4   Theorem

The image of a path connected space under a continuous mapping is path connected.

Proof Suppose that X is path connected and g: $X \to Y$ is a continuous surjective map. If a,b are two points of Y then there are two points a', b' in X with $g(a') = a$ and $g(b') = b$. Since X is path connected there is a path f from a' to b'. But then gf is a path from a to b which shows that Y is path connected.

12.5   Corollary

If X and Y are homeomorphic topological spaces then X is path connected if and only if Y is path connected.

From the theorem we deduce that S^1 is path connected. Using this we could then show that $R^{n+1} - \{0\}$, S^n and $R P^n$ for $n \geq 1$ are path connected (for $R^{n+1} - \{0\}$ because any pair of points in $R^{n+1} - \{0\}$ lie on some circle not passing through $\{0\}$, for S^n and $R P^n$ because they are continuous images of $R^{n+1} - \{0\}$).

12.6   Theorem

Suppose that $\{Y_j, j \in J\}$ is a collection of path connected subsets of a space X. If $\cap_{j \in J} Y_j \neq \emptyset$ then $Y = \cup_{j \in J} Y_j$ is path connected.

Proof Suppose that a, $b \in Y$, then $a \in Y_k$ and $b \in Y_\ell$ for some k, $\ell \in J$. Let c be any point of $\cap_{j \in J} Y_j$. Since Y_k is path connected and a, $c \in Y_k$ there is a path f from a to c. Similarly there is a path g from c to b. There is a path h from a to b given by $h = f*g$, i.e.

$$h(t) = \begin{cases} f(2t) & \text{if } 0 \leq t \leq \frac{1}{2}, \\ g(2t-1) & \text{if } \frac{1}{2} \leq t \leq 1. \end{cases}$$

The above result provides an alternative way of showing that $R^{n+1} - \{0\}$ (and hence S^n and $R P^n$) for $n \geq 1$ is path connected.

12.7   Theorem

Let X and Y be topological spaces. Then X and Y are path connected if and only if X × Y is path connected.

The proof, which is identical to the proof of Theorem 9.7 (but with the word path connected replacing the word connected), is left for the reader.

The above results should not mislead the reader into thinking that there is no difference between connectedness and path connectedness. The next result shows this.

12.8 Theorem

Every path connected space is connected. Not every connected space is path connected.

Proof Suppose that X is a path connected space. We shall prove that X is connected. To this end let $X = U \cup V$ with U,V open and non-empty. Since X is path connected and U,V are non-empty there is a path f: $[0,1] \to X$ with $f(0) \in U$ and $f(1) \in V$. Since $[0,1]$ is connected so is $f([0,1])$ and so $U \cap f([0,1])$, $V \cap f([0,1])$ cannot be disjoint. Thus neither can U and V be disjoint and so X is connected.

To show that not all connected spaces are path connected we shall give an example known as the *flea and comb* (see Figure 12.2). Consider the subset $X \subseteq \mathbb{C}$ where $X = A \cup B$ with

$A = \{ i \}$ (the flea),
$B = [0,1] \cup \{ (1/n) + y \, i; n \in \mathbb{N}, 0 \leq y \leq 1 \}$ (the comb).

We claim that X is connected but not path connected. To prove that X is connected first observe that B is path connected (use Theorem 12.6 on $B_n =$

Figure 12.2. The flea and comb.

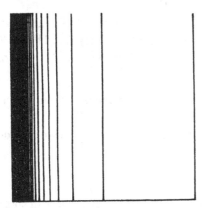

$[0,1] \cup \{ (1/n) + y \, i; 0 \leq y \leq 1 \}$ for $n \in \mathbb{N}$ and then on $B = \underset{n \in \mathbb{N}}{\cup} B_n$) and
so B is connected. Suppose that U is an open and closed subset of X. We may suppose that $A \subseteq U$ (otherwise the complement of U is an open and closed subset of X that contains A). Since U is open and $i \in U$ there is an $\epsilon > 0$ such that

$$\{ x; |i-x| < \epsilon \} \cap X \subseteq U.$$

There is some integer n such that $(1/n) + i \in U$; in particular $U \cap B \neq \emptyset$. But B is connected and $U \cap B$ is a non-empty open and closed subset of B; thus $U \cap B = B$, i.e. $B \subseteq U$. But $X = A \cup B$ and $A \subseteq U$; thus $U = X$ and so X is connected. (Essentially we have proved that $B \subseteq X \subseteq \bar{B}$, so that X being connected follows from Exercise 9.8(f).)

To prove that X is not path connected we shall show that the only path in X that begins at $i \in X$ is the constant path. Let f be the path in X that begins at $i \in X$. Since i is closed in X, $f^{-1}(i)$ is closed in $[0,1]$, furthermore $f^{-1}(i) \neq \emptyset$ since $0 \in f^{-1}(i)$. Let U be the open subset of X given by

$$U = X \cap \{ z \in \mathbb{C} ; |z-i| < \tfrac{1}{2} \}.$$

If $t_0 \in f^{-1}(i)$ then since f is continuous there is an $\epsilon > 0$ such that $f(t) \in U$ whenever $|t-t_0| < \epsilon$. We claim that $f((t_0 - \epsilon, t_0 + \epsilon) \cap [0,1]) = i$. To see this suppose that $|t_1 - t_0| < \epsilon$ and $f(t_1) \in B$. Since $U \cap B$ is a union of disjoint intervals, the interval containing $f(t_1)$ is both open and closed in U (open because U is open, and closed because the interval is of the form $\{ (1/n) + yi; 0 \leq y \leq 1 \} \cap U$ for some $n \in \mathbb{N}$). But this contradicts the fact that $f((t_0 - \epsilon, t_0 + \epsilon) \cap [0,1])$ is connected. Hence

$$(t_0 - \epsilon, t_0 + \epsilon) \cap [0,1] \subseteq f^{-1}(i).$$

We have just shown that if $t_0 \in f^{-1}(i)$ then

$$(t_0 - \epsilon, t_0 + \epsilon) \cap [0,1] \subseteq f^{-1}(i),$$

which means that $f^{-1}(i)$ is open. Since $f^{-1}(i)$ is also closed and $[0,1]$ is connected it follows that $f^{-1}(i) = [0,1]$, in other words $f([0,1]) = i$. There is therefore no path between $i \in X$ and any point $B \subseteq X$; thus X is not path connected as was claimed.

There are many other (similar) examples of connected but not path connected spaces; see the exercises below.

The last result that we prove in this chapter concerns open connected subsets of $\mathbb{R}^n$.

12.9 Theorem
Any non-empty open connected subset E of $\mathbb{R}^n$ is path connected.

Proof Let $p \in E$ and let F be the subset of E that consists of those points in E that can be joined to p by a path in E. We claim that E is open. To prove this let $q \in F \subseteq E$. Since E is open there is an open n-disc $D \subseteq E$ with centre q, i.e.

$$q \in D = \{ x; \|q - x\| < \epsilon \} \subseteq E$$

for some $\epsilon > 0$. The open n-disc D is path connected (it is homeomorphic to $\mathbb{R}^n$) hence any point of D can be joined to q by a path in D. Hence any point of D can be joined to p by a path in E and so $q \in D \subseteq F$. Thus F is open.

We also claim that F is closed. To see this consider $G = E - F$; thus G consists of those points in E that cannot be joined to p by a path in E. By an argument similar to the one above we can show that G is open and hence F is closed. The subset F is non-empty, open and closed; since E is connected $F = E$ and so E is path connected.

12.10 Exercises

(a) Prove that any space with the concrete topology is path connected.

(b) Which of the following subsets of $\mathbb{C}$ are path connected?

$$\{ z; |z| \neq 1 \} , \{ z; |z| \geq 1 \} , \{ z; z^2 \text{ is real} \} .$$

(c) Prove the result in Lemma 12.2 but with the assumption that A and B are both open subsets of W.

(d) Let $X = A \cup B$ be the subspace of $\mathbb{R}^2$ where

$A = \{ (x,y); x = 0, -1 \leq y \leq 1 \} ,$
$B = \{ (x,y); 0 < x \leq 1, y = \cos(\pi/x) \} .$

Show that X is connected but not path connected.

(e) Let $X = A \cup B$ be the subspace of $\mathbb{R}^2$ where

$A = \{ (x,y); x = 0, -1 \leq y \leq 1 \}$
$B = \{ (x,y); 0 < x \leq 1, y = \sin(1/x) \}$

Show that X is connected but not path connected.

(f) Consider the following subsets of $\mathbb{R}^2$
$A = \{ (x,y); 0 \leq x \leq 1, y = x/n \text{ for } n \in \mathbb{N} \}$
$B = \{ (x,y); \frac{1}{2} \leq x \leq 1, y = 0 \}$

Prove that $A \cup B$ is connected but not path connected.

(g) Suppose that A is a path connected subset of a space X and that $\{ A_j; j \in J \}$ is a collection of path connected subsets of X each of which intersects A. Prove that $A \cup \{ \underset{j \in J}{\cup} A_j \}$ is path connected.

(h) Let $S^n = S^n_+ \cup S^n_-$ where

$S^n_+ = \{\, x \in \mathbb{R}^{n+1}; \|x\| = 1, x_{n+1} \geq 0 \,\}$,
$S^n_- = \{\, x \in \mathbb{R}^{n+1}; \|x\| = 1, x_{n+1} \leq 0 \,\}$.

Using Exercise 8.14(h) prove that S^n is path connected for $n > 0$.

(i) Let $\sim$ be the relation on the points of a space X defined by saying that $x \sim y$ if and only if there is a path in X joining x and y. Prove that $\sim$ is an equivalence relation. Prove also that X is path connected if and only if the quotient space $X/\sim$ is path connected.

(j) An open neighbourhood of a point $x \in X$ is an open set U such that $x \in U$. A space X is said to be *locally path connected* if for all $x \in X$ every open neighbourhood of x contains a path connected open neighbourhood of x. Prove that if X is locally path connected and $U \subseteq X$ is open in X then U is locally path connected. Prove that $\mathbb{R}^n$ is locally path connected (and hence every open subset of $\mathbb{R}^n$ is locally path connected). Prove that if X is locally path connected and connected then X is path connected (this therefore reproves Theorem 12.9).

(k) Let $p, q \in X$. The subsets $A_1, A_2, ..., A_k$ of X are said to form a *simple chain* joining p and q if $p \in A_1$, $q \in A_k$, $A_i \cap A_j = \emptyset$ whenever $|i-j| > 1$ and $A_i \cap A_{i+1} \neq \emptyset$ for $i = 1,2,...,k-1$; see Figure 12.3.

Figure 12.3

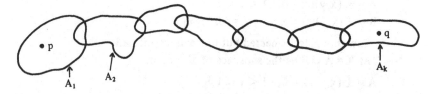

Prove that if X is connected and if $\{\, U_j; j \in J \,\}$ is an open cover of X then any pair of points in X can be joined by a simple chain consisting of members of $\{\, U_j; j \in J \,\}$. (Hint: For $p \in X$ consider the set of points in X which can be joined to p by some simple chain consisting of members of $\{\, U_j; j \in J \,\}$.)

(l) Use (k) above to give yet another proof of Theorem 12.9.

(m) Prove that a connected n-manifold is path connected.

(n) Prove that an n-manifold is locally path connected.

(o) Prove that the space $Y \subseteq \mathbb{R}^2$ given by $Y = A \cup B \cup C$ where

$A = \{\, (x,y); x^2 + y^2 = 1, y \geq 0 \,\}$,
$B = \{\, (x,y); -1 \leq x \leq 0, y = 0 \,\}$,

$C = \{ (x,y); 0 < x \le 1, y = \tfrac{1}{2} \sin (\pi/x) \}$.

is path connected but not locally path connected.

(p) Let $Z = Y \cup D \in \mathbb{R}^2$ where Y is as above in (o) and D is the circle $\{ (x-1)^2 + y^2 = 1 \}$. Prove that Z is path connected but not locally path connected.

We end this chapter with a strange path. This is a path $f: I \to I^2$ which is surjective. Such examples are referred to as 'space filling curves'. They were first invented by G. Peano in about 1890. The path f is defined as the limit of paths $f_n: I \to I^2$. The first three are illustrated in Figure 12.4. The reader should have no trouble in visualizing the n-th step. After n steps every point of the square I^2 lies within a distance of $(\tfrac{1}{2})^n$ of a point in $f_n(I)$. In the limit we get a continuous surjective map $f: I \to I^2$. Note that at any finite stage the continuous map $f_n: I \to I^2$ fails to be injective only at $\{0\}$ and $\{1\}$ in I. In fact $\overset{\circ}{I}$ and $f_n(\overset{\circ}{I})$ are homeomorphic. This is certainly not true in the limit.

Figure 12.4

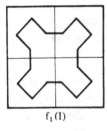 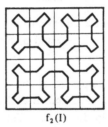 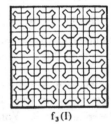

$f_1(I)$ $f_2(I)$ $f_3(I)$

12A.1   Definition

A *simple closed curve* C is a homeomorphic image of the circle. A *component* is a maximal connected subspace.

One of the two following statements is true and one is false.

(A) Let C be a simple closed curve in the euclidean plane. Then $\mathbb{R}^2 - C$ is disconnected and consists of two components with C as their common boundary. Exactly one of these components is bounded.

(B) Let D be a subset of the euclidean plane. If D is the boundary of each component of its complement $\mathbb{R}^2 - D$ and if $\mathbb{R}^2 - D$ has a bounded component then D is a simple closed curve.

We now construct an example to show that not both of the above statements are true. The example is known as the *Lakes of Wada*. It was first described by K. Yoneyama in 1917. Consider a region in the form of a double annulus; see Figure 12A.1. We imagine this to be an island, surrounded by sea water, and having two lakes. For convenience the water in the two lakes consists of different colours. By constructing canals from the sea and the lakes into the island we shall define three connected open sets. At time $t=0$ we construct a canal from the sea bringing sea water to within a distance of 1 unit of every point of land. At time $t=\frac{1}{2}$ we dig a canal from lake 1 bringing water from that lake to within a distance $\frac{1}{2}$ of every point of land. At time $t=\frac{3}{4}$ we dig a canal from lake 2 to bring water from that lake to within a distance $\frac{1}{4}$ of every point of land. The process continues where we build canals at time $1-(\frac{1}{2})^n$ bringing the appropriate water to within a distance $(\frac{1}{2})^n$ of every point of land. The canals of course must have no intersections. The two lakes with their canal systems and the sea with its canal form three open sets, each connected, with the remaining 'dry land' D as common boundary.

Now if (B) is true then the region D in the Lakes of Wada is a simple closed curve, and hence (A) is false. On the other hand if (A) is true then (B) must be false. This proves our first assertion that at most one of the statements (A), (B) is true.

In fact (A) is true while (B) is false. The result (A) is called the *Jordan curve theorem*. It is named after C. Jordan, who pointed out in the early 1890s that although (A) may seem intuitively obvious a rigorous proof is required. Such a proof was given in the early 1900s by O. Veblen. The proof that we give here is based upon a recently discovered 'elementary' proof by Helge Tverberg to whom we are indebted.

For simplicity we shall call a simple closed curve in the plane a *Jordan curve*. Thus a Jordan curve C is a subspace of $\mathbb{R}^2$ homeomorphic to $S^1 = \{ z \in \mathbb{C} ; |z| = 1 \}$. We shall say that a Jordan curve C is *given* by f: $S^1 \rightarrow \mathbb{R}^2$ if C = f(S^1); of course f is not unique. A Jordan curve is a *Jordan polygon* if it consists of a finite number of straight-line segments.

We always think of the circle S^1 as a subset of the complex plane and it

Figure 12A.1. The Lakes of Wada.

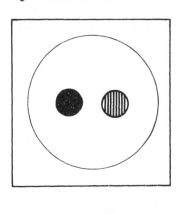
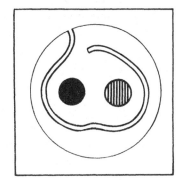

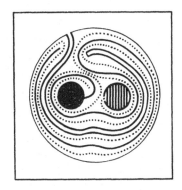

is convenient to think of $\mathbb{R}^2$ as the complex plane. Thus the distance between two points x,y of $\mathbb{R}^2$ or of S^1 will be denoted by |x−y|. If A and B are two disjoint compact subsets then we define d(A,B) by

$$d(A,B) = \inf \{ |a - b| ; a \in A, b \in B \}.$$

In particular if A consists of a single point, say { x }, we have

$$d(x,B) = \inf \{ |x - b| ; b \in B \}.$$

Our first result is to show that the Jordan curve theorem holds for a Jordan polygon.

12A.2   Theorem

The Jordan curve theorem holds for a Jordan polygon, i.e. if C is a Jordan polygon then $\mathbb{R}^2$−C consists of two components with C as the common boundary and exactly one of these components is bounded.

Proof First we shall show that if C is a Jordan polygon then $\mathbb{R}^2$−C has at least two components. Let p ∈ $\mathbb{R}^2$−C and consider any straight line r beginning at p; call such a line a *ray* at p. Let P(r,p) denote the number of times that r intersects C with the convention that if r passes through a vertex

Figure 12A.2

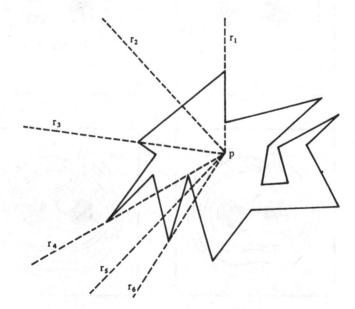

V or intersects a whole-line segment L of C then we count such an intersection as two if both adjacent edges to V or L are on the same side of V or L respectively; otherwise we count it as one. For example in Figure 12A.2 we have $P(r_1,p) = 1$, $P(r_2,p) = 1$, $P(r_3,p) = 1$, $P(r_4,p) = 5$, $P(r_5,p) = 3$ and $P(r_6,p) = 3$.

As the ray r at p rotates, the value of $P(r,p)$ in general changes, but whether it is even or odd does not. We therefore define p to be an *even* or *odd* point according to whether $P(r,p)$ is even or odd respectively for any ray r at p. We also refer to the *parity* of p being even or odd.

Thus $\mathbb{R}^2 - C$ is divided into even and odd points, X_e and X_o respectively. Clearly $\mathbb{R}^2 - C = X_e \cup X_o$ and $X_e \cap X_o = \emptyset$. We shall show that both X_e and X_o are open subsets of $\mathbb{R}^2 - C$. Let $p \in \mathbb{R}^2 - C$ and suppose that $d(p,C) = \epsilon$. This means that $B_\epsilon(p) \subseteq \mathbb{R}^2 - C$. The parity of all points in $B_\epsilon(p)$ is the same as the parity of p; for $x \in B_\epsilon(p)$ consider the ray at p passing through x. Thus X_e and X_o are open so that $\mathbb{R}^2 - C$ is disconnected and consists of at least two components.

Both X_e and X_o are path connected. To see this choose any straight-line segment in C and let a,b be two points in $\mathbb{R}^2 - C$, close to C but on opposite sides of this line segment so that $a \in X_e$ and $b \in X_o$. Now, if p is any point in $\mathbb{R}^2 - C$ then there is clearly a path in $\mathbb{R}^2 - C$ that goes from p to a point close to C. By continuing this path, remaining in $\mathbb{R}^2 - C$ and close to C, we eventually reach a or b. This shows that X_e and X_o are path connected and hence connected, which completes the proof of the theorem.

To continue we shall need the concept of 'uniform continuity' and the fact that if f: $S^1 \to \mathbb{R}^2$ is a continuous map then f is uniformly continuous.

12A.3 Definition

Let M_1, M_2 be metric spaces with metrics d_1, d_2 respectively. A map f: $M_1 \to M_2$ is *uniformly continuous* if given $\epsilon > 0$ there exists $\delta > 0$ such that $d_2(f(x), f(y)) < \epsilon$ for all x,y in M_1 satisfying $d_1(x,y) < \delta$.

Note that this is stronger than ordinary continuity.

12A.4 Theorem

Let M_1, M_2 be metric spaces with metrics d_1, d_2 respectively. If f: $M_1 \to M_2$ is a continuous map and if M_1 is compact then f is uniformly continuous.

Proof Let $\epsilon > 0$. For all $x \in M_1$ there exists $\delta(x) > 0$ such that if $y \in M_1$ and $d_1(x,y) < 2\delta(X)$ then $d_2(f(x), f(y)) < \tfrac{1}{2}\epsilon$. The set

$$\{ B_{\delta(x)}(x) ; x \in M_1 \}$$

is an open cover of M_1. Since M_1 is compact there is a finite subcover:

$$\{ B_{\delta(x_1)}(x_1), B_{\delta(x_2)}(x_2), ..., B_{\delta(x_n)}(x_n) \} .$$

Let $\delta = \min \{ \delta(x_1), \delta(x_2),..., \delta(x_n) \}$. If $x,y \in M_1$ and $d_1(x,y) < \delta$ then $x \in B_{\delta(x_i)}(x_i)$ for some i $(1 \leq i \leq n)$ and so

$$d_2(f(x), f(x_i)) < \tfrac{1}{2} \epsilon$$

since $\delta \leq \delta(x_i)$. Also

$$d_1(y,x_i) \leq d_1(y,x) + d_1(x,x_1) < \delta + \delta(x_i) \leq 2\delta(x_i)$$

so that

$$d_2(f(y), f(x_i)) \leq \tfrac{1}{2} \epsilon.$$

Thus

$$d_2(f(x), f(y)) \leq d_2(f(x), f(x_i)) + d_2(f(x_i), f(y)) < \epsilon$$

which proves the result.

12A.5   Corollary
If f: $S^1 \to \mathbb{R}^2$ is a continuous map then f is uniformly continuous.

The proof is obvious.

12A.6   Corollary
Let M_1, M_2 be metric spaces with metrics d_1, d_2 respectively. If f: $M_1 \to M_2$ is a continuous map with M_1 compact and with f: $M_1 \to f(M_1)$ a homeomorphism then, given $\epsilon > 0$, there exists $\delta > 0$ such that $d_1(x,y) < \epsilon$ whenever $d_2(f(x), f(y)) < \delta$.

Proof f^{-1}: $f(M_1) \to M_1$ is a continuous map between metric spaces with $f(M_1)$ compact.

12A.7   Theorem
Let C be a Jordan curve given by f: $S^1 \to \mathbb{R}^2$. For every $\epsilon > 0$ there exists a Jordan polygon C' given by f': $S^1 \to \mathbb{R}^2$ such that $|f(x) - f'(x)| < \epsilon$ for all $x \in S^1$.

Proof Since f is uniformly continuous on S^1 there exists $\epsilon_1 > 0$ such that

$$|x-y| < \epsilon_1 \Rightarrow |f(x) - f(y)| < \tfrac{1}{2} \epsilon.$$

Since f: $S^1 \to C$ is a homeomorphism we have by Corollary 12A.6 that there exists $\epsilon_2 > 0$ such that

$$|f(x) - f(y)| < \epsilon_2 \Rightarrow |x-y| < \min (\epsilon_1, \sqrt{3})$$

(The reason for the $\sqrt{3}$ is that if A is a subset of S^1 of diameter less than $\sqrt{3}$ then A is contained in a smallest closed arc.)

Let $\delta = \min(\tfrac{1}{2} \epsilon, \epsilon_2)$. Cover C by square regions $S_1, S_2, ..., S_n$ that do not overlap (except at the edges), with each square being of diameter δ. Because $\delta \leq \epsilon_2$ we know that $f^{-1}(S_1)$ is contained in a smallest closed arc $A_1 \neq S^1$. Now straighten $f(A_1)$ to form a Jordan curve C_1; in other words define $f_1 : S^1 \rightarrow \mathbb{R}^2$ by

$$f_1(e(t)) = \begin{cases} f(e(t)), & \text{if } e(t) \notin A_1, \\ \left(1 - \dfrac{t-a}{b-a}\right) f(e(a)) + \dfrac{t-a}{b-a} f(e(b)) & \text{if } e(t) \in A_1 \end{cases}$$

where $A_1 = \{ e(t); a \leq t \leq b \}$ and $e(t) = \exp(2\pi i t)$, then let $C_1 = f_1(S^1)$ which is clearly a Jordan curve. Note that $f(A_1)$ is not necessarily contained in S_1. Note also that $f_1^{-1}(S_i) \subseteq f^{-1}(S_i)$ for i=2,3,...,n.

Next straighten $f_1(A_2)$ where A_2 is the smallest arc containing $f_1^{-1}(S_2)$. This gives us a Jordan curve C_2 given by $f_2 : S^1 \rightarrow \mathbb{R}^2$ (if $f_1^{-1}(S_2) = \emptyset$ then put $f_2 = f_1$ and $C_2 = C_1$). Note again that $f_2^{-1}(S_i) \subseteq f^{-1}(S_i)$ for i=3,4,...,n. Continuing in this way we obtain a Jordan polygon C_n given by $f_n : S^1 \rightarrow \mathbb{R}^2$. We shall check that C_n is ϵ-close to C.

Suppose $x \in S^1$ and $f_n(x) \neq f(x)$. Then $f_n(x) = f_j(x) \neq f_{j-1}(x)$ for some $j \geq 1$ where $f_0 = f$. By construction x belongs to an arc A_j with end points y,z say. Also, by construction, $f_j(y) = f(y)$ and $f_j(z) = f(z)$. We have

$$\begin{aligned} |f(x) - f_n(x)| &= |f(x) - f(y) + f_j(y) - f_j(x)| \\ &\leq |f(x) - f(y)| + |f_j(y) - f_j(x)| \\ &\leq |f(x) - f(y)| + \delta \\ &\leq |f(x) - f(y)| + \tfrac{1}{2} \epsilon. \end{aligned}$$

Since $|f(z) - f(y)| \leq \delta \leq \epsilon_2$ we deduce that $|z-y| \leq \epsilon_1$. But $|x-y| < |z-y|$, so that $|x-y| < \epsilon_1$ and hence $|f(x) - f(y)| < \tfrac{1}{2} \epsilon$. Thus

$$|f(x) - f_n(x)| < \tfrac{1}{2} \epsilon + \tfrac{1}{2} \epsilon = \epsilon.$$

12A.8 Theorem

Let C be a Jordan polygon given by f: $S^1 \rightarrow \mathbb{R}^2$. Then the bounded component of $\mathbb{R}^2 - C$ contains an open disc whose boundary circle meets C at two points f(a), f(b) with $|a-b| \geq \sqrt{3}$.

Proof Let D be an open disc such that $D \subseteq \mathbb{R}^2 - C$ and such that there exists two points f(a), f(b) $\in \partial D$ with $|a-b|$ maximal. Such a disc exists. Suppose that $|a-b| < \sqrt{3}$. Then a and b must be the end points of an arc A of length greater than $4\pi/3$. The boundary circle of D cannot meet $f(A)$ – $\{ f(a), f(b) \}$ because max $\{ |a-c|, |b-c| \} > |a-b|$ for all $c \in A - \{ a,b \}$.

Let $f(v_1)$, $f(v_2)$,...,$f(v_n)$ be the vertices of C in $f(A)$ as seen when going from $f(a)$ to $f(b)$. Four possibilities could arise: (i) $v_1 \neq a$, $v_n \neq b$, (ii) $v_1 \neq a$, $v_n = b$, (iii) $v_1 = a$, $v_n \neq b$ and (iv) $v_1 = a$, $v_n = b$. In the first case the circle ∂D is tangent to the line segments $f(a) f(v_1)$ and $f(b) f(v_n)$. There is a disc $D' \subseteq R^2-C$ close to D such that the circle $\partial D'$ touches C at points close to $f(a)$ and $f(b)$, say at $f(a')$ and $f(b')$, which lie in the line segments $f(a) f(v_1)$ and $f(b) f(v_n)$ respectively; see Figure 12A.3(a). As $|a'-b'| > |a-b|$ we get a contradiction. In case (ii) the circle ∂D is tangent to $f(a) f(v_1)$ and there is a disc $D' \subseteq R^2-C$ such that $\partial D'$ touches C close to $f(a)$ in $f(a) f(v_1)$ and passes through $f(b) f(v_{n-1})$; see Figure 12A.3(b). This leads to a contradiction. Case (iii) is similar to (ii). For case (iv) consider the region R bounded by $f(A)$ and the radii of $\bar{D}$ to $f(a)$ and $f(b)$. For each $x \in R$ there is a unique circle S_x that has centre x and passes through $f(a)$ and $f(b)$. Letting x move continuously from the centre of D we get, for some x, circles S_x that bound discs D_x with $D_x \subseteq R^2-C$. Eventually, for some particular x, the circle S_x either meets $f(A)$ at some point other than $f(a)$ and $f(b)$, or becomes tangent to one of the two segments $f(a) f(v_2)$, $f(b) f(v_{n-1})$. The first case has already been shown to be impossible while the second gives a contradiction by the method of case (ii) above. All these contradictions leave only one possibility, namely that $|a-b| \geq \sqrt{3}$.

We now prove part of the Jordan curve theorem.

Figure 12A.3

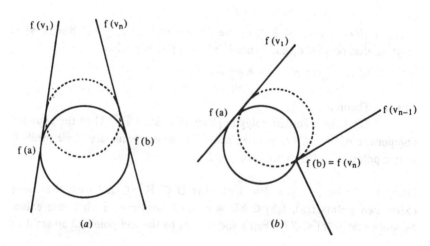

$f(v_1)$ $f(v_n)$

$f(a)$ $f(b)$

(a)

$f(v_1)$

$f(a)$ $f(v_{n-1})$

$f(b) = f(v_n)$

(b)

12A.9 Theorem

If C is a Jordan curve then $\mathbb{R}^2 - C$ has at least two components.

Proof There is clearly an unbounded component. We shall show that there is also a bounded component. Let $C_1, C_2, \ldots$ be a sequence of Jordan polygons converging to C, (as determined in Theorem 12A.7 with ϵ being $\epsilon_1, \epsilon_2, \ldots$ where the ϵ_n converge to 0). Let $C, C_1, C_2, \ldots$ be given by $f, f_1, f_2, \ldots$ respectively, so that f_n converges to f as n goes to infinity. For each C_n there is a circle S_n containing points $f_n(a_n)$ and $f_n(b_n)$ with $|a_n - b_n| \geq \sqrt{3}$, by Theorem 12A.8. Let the centre of S_n be z_n. There is a circle S_0 that surrounds all the Jordan curves C_n and C, hence S_0 surrounds all the S_n. Thus the sequence $z_1, z_2, \ldots$ is a bounded sequence in $\mathbb{R}^2$ and hence has a convergent subsequence. We may therefore assume that the sequence z_n, $n \geq 1$, converges to z as n goes to infinity.

For large n the points z and z_n lie in the same component of $\mathbb{R}^2 - C_n$. We see this as follows: There exists $\delta > 0$ such that if $|x-y| \geq \sqrt{3}$ then $|f(x) - f(y)| \geq \delta$. Thus $|f(a_n) - f(b_n)| \geq \delta$ for all $n \geq 1$ and hence $|f_n(a_n) - f_n(b_n)| > \tfrac{1}{2}\delta$ for $n \geq N$, where N is sufficiently large so that $\epsilon_N < \tfrac{1}{2}\delta$. This means that the diameter of S_n, for $n \geq N$, is greater than $\tfrac{1}{2}\delta$ and so $d(z_n, C_n) > \tfrac{1}{4}\delta$. But for large enough n we have $|z - z_n| < \tfrac{1}{4}\delta$, so that z and z_n must lie in the same component of $\mathbb{R}^2 - C_n$, namely the bounded component of $\mathbb{R}^2 - C_n$, since by definition z_n lies in the bounded component of $\mathbb{R}^2 - C_n$. We shall show that z cannot lie in the unbounded component of $\mathbb{R}^2 - C$.

Suppose that z lies in the unbounded component of $\mathbb{R}^2 - C$. There is therefore a continuous path g in $\mathbb{R}^2 - C$ from z to some point outside C_0 (by Theorem 12.9, an open connected subset of $\mathbb{R}^2$ is path connected). Let $d(g(I), C) = \delta$. For large n we have $|f_n(x) - f(x)| < \tfrac{1}{2}\delta$ and hence $d(g(I), C_n) > \tfrac{1}{2}\delta$, which means that for large n the point z is in the unbounded component of $\mathbb{R}^2 - C_n$. But this contradicts what was shown previously, namely that z lies in the bounded component of $\mathbb{R}^2 - C_n$. We conclude therefore that z does not lie in the unbounded component of $\mathbb{R}^2 - C$ and so $\mathbb{R}^2 - C$ has also a bounded component.

To prove the second part of the Jordan curve theorem we need a definition and a lemma.

12A.10 Definition

A *chord* Γ of a Jordan curve C is a straight-line segment that intersects C only at its end points. Thus, apart from the end points of Γ, Γ lies in $\mathbb{R}^2 - C$.

Note that if C is a Jordan polygon and Γ is a chord of C then $\Gamma \subseteq X \cup C$ where X is one of the components of $\mathbb{R}^2-C$, and furthermore $X-\Gamma$ consists of two components.

12A.11 Lemma
Let C be a Jordan polygon and let a,b be two points in the same component X of $\mathbb{R}^2-C$ such that $d(\{a,b\},C) \geq \delta$, some $\delta > 0$. Suppose that whenever Γ is a chord of C in $X \cup C$ of length less than 2δ then both a and b are in the same component of $X-\Gamma$. In such a situation there is a path g in X such that $d(g(I),C) \geq \delta$.

Proof The idea is to place an open disc of radius δ at centre a and then to pull it towards b keeping it within X. The only way that we might be prevented from pulling the disc (of diameter 2δ) is if there is a chord of length less than 2δ in $X \cup C$. The assumptions on the chords ensure that this cannot happen.

12A.12 Theorem
Let C be a Jordan curve; then $\mathbb{R}^2-C$ has at most two components.

Proof Suppose that $\mathbb{R}^2-C$ has three or more components and let p,q,r be points from three distinct components. Let $d(\{p,q,r\},C) = \epsilon$ and let $C_1,C_2,...$ be a sequence of Jordan polygons converging to C. Suppose that $C,C_1,C_2,...$ are given by $f,f_1,f_2,...$ respectively. For n large $d(C_n,C) < \frac{1}{2}\epsilon$ and so $d(\{p,q,r\},C_n) > \frac{1}{2}\epsilon$. Using Theorem 12A.2 we see that, for each n sufficiently large, two of the three points p,q,r are in the same component X_n of $\mathbb{R}^2-C_n$. By passing to a subsequence, if necessary, we may assume that p and q are in X_n for all n.

Suppose that there is a δ with $0 < \delta < \epsilon$ and infinitely many n such that the points p and q are connected by a path g_n in X_n with $d(g_n(I),C_n) \geq \delta$. For large n we have $d(C_n,C) < \frac{1}{2}\delta$ and so $d(g_n(I),C) > \frac{1}{2}\delta$ for n large, which shows that p and q are in the same component of $\mathbb{R}^2-C$. This contradiction means that no such δ exists. By using Lemma 12A.11 we deduce that for infinitely many n there exists a chord Γ_n of length δ_n with the properties that p and q are in different components of $X_n-\Gamma_n$ and δ_n tends to 0 as n tends to infinity. Let these infinitely many n be denoted in increasing order by $n(1),n(2),...$ Also, let the end points of $\Gamma_{n(i)}$ be $f_{n(i)}(a_i)$, $f_{n(i)}(b_i)$. Since $\delta_{n(i)} \to 0$ as $i \to \infty$ we have

$$f_{n(i)}(a_i) - f_{n(i)}(b_i) \to 0 \text{ as } i \to \infty,$$

and hence

$f(a_i) - f(b_i) \to 0$ as $i \to \infty$,

which implies that

$a_i - b_i \to 0$ as $i \to \infty$.

Since p and q are in different components of $X_{n(i)} - \Gamma_{n(i)}$ then for infinitely many values of i one of the points, say p, belongs to the component of $X_{n(i)} - \Gamma_{n(i)}$ bounded by $\Gamma_{n(i)}$ and $f_{n(i)}(A_i)$, where A_i is the smallest of the arcs on S^1 with end points a_i, b_i. Since $a_i - b_i \to 0$ as $i \to \infty$ it follows that the diameter of this component just defined is smaller than ϵ, for i sufficiently large. In particular $|p - f(a_i)| < \epsilon$, which is a contradiction, proving the theorem.

The Jordan curve theorem follows from Theorems 12A.9 to 12A.12.

12A.13 Exercises

(a) Prove that if A is the image in $\mathbb{R}^2$ of an injective continuous map f: $I \to \mathbb{R}^2$ then $\mathbb{R}^2 - A$ is connected.

(b) Let C be a Jordan curve given by f: $S^1 \to \mathbb{R}^2$. Define δ by

$\delta = \min \{ |f(x) - f(y)|; x, y \in S^1 \text{ with } |x-y| \geq \sqrt{3} \}$.

Prove that the bounded component of $\mathbb{R}^2 - C$ contains an open disc of diameter δ.

(c) It is possible to fit an uncountable number of simple closed curves disjointly in $\mathbb{R}^2$; for example $\{ C_r; r \in \mathbb{R} \}$ where $C_r = \{ (x,y) \in \mathbb{R}^2; x^2 + y^2 = r \}$. A *figure eight curve* is a space homeomorphic to

$\{ (x,y) \in \mathbb{R}^2; (x\pm1)^2 + y^2 = 1 \}$.

Prove that if $\{ E_j; j \in J \}$ is a disjoint collection of figure eight curves in $\mathbb{R}^2$ then J is countable.

13

Homotopy of continuous mappings

In this chapter we introduce equivalence relations on continuous mappings between topological spaces. This will be of fundamental importance in subsequent chapters, particularly when applied to paths.

Roughly speaking two continuous maps f_0, f_1: $X \to Y$ are said to be homotopic if there is an intermediate family of continuous maps f_t: $X \to Y$, for $0 \leq t \leq 1$ which vary continuously with respect to t. See Figure 13.1(a). Figure 13.1(b) depicts two maps that are not homotopic; here $X = S^1$ and Y is an annulus in $\mathbb{R}^2$.

Figure 13.1. (a) Homotopic maps. (b) Non-homotopic maps.

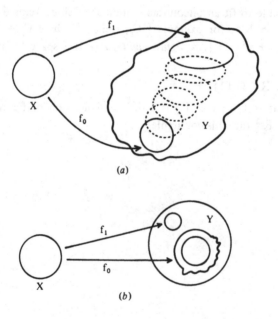

(a)

(b)

More precisely we have

13.1   Definition

Two continuous maps f_0, f_1: $X \to Y$ are said to be *homotopic* if there is a continuous map F: $X \times I \to Y$ such that $F(x,0) = f_0(x)$ and $F(x,1) = f_1(x)$.

For an example see Figure 13.2.

Figure 13.2

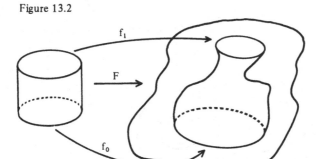

The map F is called a *homotopy* between f_0 and f_1. We write $f_0 \simeq f_1$ or F: $f_0 \simeq f_1$. For each $t \in [0,1]$, we denote $F(x,t)$ by $f_t(x)$, whence f_t: $X \to Y$ is a continuous map.

Observe that if f: $I \to Y$ is a path then f is homotopic to the constant path $\epsilon_{f(0)}$ by the homotopy F: $I \times I \to Y$ where $F(x,t) = f((1-t)x)$. To avoid such situations (that is, if we want to) we use a more general concept of homotopy – that of homotopy relative to a subset A. Here we require that the homotopy does not move any point of A.

13.2   Definition

Suppose that A is a subset of X and that f_0, f_1 are two continuous maps from X to Y. We say that f_0 and f_1 are *homotopic relative* to A if there is a homotopy F: $X \times I \to Y$ between f_0 and f_1 such that $F(a,t)$ does not depend on t for $a \in A$; in other words $F(a,t) = f_0(a)$ for all $a \in A$ and all $t \in I$.

Note then that $f_0(a) = f_1(a)$ for all $a \in A$. The homotopy F is then called a *homotopy relative* to A and we write $f_0 \simeq f_1$ (rel A) or $f_0 \simeq_{\text{rel } A} f_1$.

For example see Figure 13.3 where $X = I$ and $A = \{ 0 \} \in X$.

As another example consider Figure 13.4 where $X = I$, $A = \{ 0,1 \}$ and Y is an annulus in R^2. The maps f_0, f_1 are not homotopic relative to A although they are homotopic.

Figure 13.3

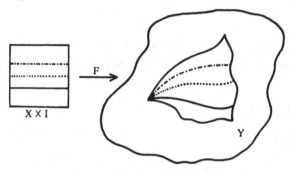

Figure 13.4

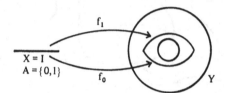

Of course, if $A = \emptyset$ then 'homotopy relative to A' becomes just 'homotopy'. The next result is that homotopy relative to A is an equivalence relation.

13.3   Lemma

The relation $\simeq_{\text{rel } A}$ on the set of continuous maps from X to Y is an equivalence relation.

Proof The relation is reflexive because $F(x,t) = f(x)$ is a homotopy relative to A between f and f itself. It is symmetric because if $F: f \simeq_{\text{rel } A} g$ then $G: g \simeq_{\text{rel } A} f$ where G is given by $G(x,t) = F(x, 1-t)$. Finally the relation is transitive because if $F: f \simeq_{\text{rel } A} g$ and $G: g \simeq_{\text{rel } A} h$ then $H: f \simeq_{\text{rel } A} h$ where H is given by

$$H(x,t) = \begin{cases} F(x,2t) & 0 \leq t \leq \tfrac{1}{2}, \\ G(x,2t-1) & \tfrac{1}{2} \leq t \leq 1. \end{cases}$$

The glueing lemma shows that H is continuous.

13.4   Exercises

(a) Let X be a space, f: $S^1 \to X$ a continuous map. Show that f is null homotopic (i.e. homotopic to a constant map) if and only if there is a continuous map g: $D^2 \to X$ with $g|S^1 = f$. (Hint: If c is a constant map and F: $c \simeq f$ then define $g(rx) = F(x,r)$ for $x \in S^1$, $r \in I$ and use Exercise 8.14(f).)

(b) Let x, y $\in$ X. Denote by P(x,y) the set of equivalence classes of paths in X from x to y under the equivalence relation 'homotopic relative to $\{\,0,1\,\}$'. (In other words two paths p,q: $I \to X$ from x to y are equivalent in P(x,y) if and only if $p \simeq q$ (rel $\{\,0,1\,\}$).) Show that there is a one-to-one correspondence between P(x,y) and P(x,x) if and only if P(x,y) $\neq \emptyset$.

(c) Let $0 < s < 1$. Given paths p and q with $p(1) = q(0)$, define h by the formula

$$h(t) = \begin{cases} p(t/s) & 0 \leq t \leq s, \\ \\ q((t-s)/(1-s)) & s \leq t \leq 1. \end{cases}$$

Prove that h and $p*q$ are homotopic relative to $\{\,0,1\,\}$.

(d) For a path f let $\bar{f}$ be the path given by $\bar{f}(t) = f(1-t)$. Prove that $f \simeq g$ (rel $\{\,0,1\,\}$) if and only if $\bar{f} \simeq \bar{g}$ (rel $\{\,0,1\,\}$).

(e) Show that if $f_0 \simeq_{\text{rel A}} f_1 \colon X \to Y$ and g: $Y \to Z$ is a continuous map then $gf_0 \simeq_{\text{rel A}} gf_1 \colon X \to Z$.

(f) Suppose that $f_0 \simeq f_1 \colon X \to Y$ and $g_0 \simeq g_1 \colon Y \to Z$. Prove that $g_0 f_0 \simeq g_1 f_1 \colon X \to Z$. (Hint: First use (e) to show that $g_0 f_0 \simeq g_0 f_1$ and then show that $g_0 f_1 \simeq g_1 f_1$.)

(g) Let X, Y be topological spaces and let $\mathcal{F}(X,Y)$ be the set of continuous functions from X to Y with the compact-open topology, (see Exercise 7.13(d)). Prove that if $f \simeq g$: $X \to Y$ then there is a path from f to g in the space $\mathcal{F}(X,Y)$. Suppose that X is compact and Hausdorff; prove that there is a path from f to g in $\mathcal{F}(X,Y)$ if and only if $f \simeq g$: $X \to Y$. (For this last result it is sufficient that X is locally compact and Hausdorff.)

We can use the concept of homotopic maps to give us an equivalence relation on topological spaces.

13.5   Definition

Two spaces X and Y are of the *same homotopy type* if there exist continuous maps f: $X \to Y$, g: $Y \to X$ such that

$$gf \simeq 1: X \to X,$$
$$fg \simeq 1: Y \to Y.$$

The maps f and g are then called *homotopy equivalences*. We also say that X and Y are *homotopy equivalent*.

Obviously, homeomorphic spaces are of the same homotopy type, but the converse is not true. For example, if $n > 0$ then the n-disc $D^n \subseteq \mathbb{R}^n$ is not homeomorphic to a single point, say $\{ y \} \subseteq D^n$, but it is of the same homotopy type as a single point. To see that they are homotopy equivalent consider the inclusion map $f: \{ y \} \to D^n$ (given by $f(y) = y$) and the constant map $g: D^n \to \{ y \}$. Clearly $gf = 1$, whereas $F: D^n \times I \to D^n$ defined by $F(x,t) = tx + (1-t)y$ is a homotopy between fg and $1: D^n \to D^n$. Spaces that are homotopy equivalent to a point are given a special name.

13.6   Definition
  A space X is said to be *contractible* if it is homotopy equivalent to a point.

Thus D^n is contractible. More generally any convex subset of $\mathbb{R}^n$ is contractible. Intuitively a space is contractible if it can be deformed within itself to a point (a circle cannot be deformed within itself to a point).

Another example of a pair of homotopy equivalent spaces is provided by the cylinder C and the circle S^1. To see this write C and S^1 as

$$C = \{ (x,y,z) \in \mathbb{R}^3 ; x^2 + y^2 = 1, -1 \leq z \leq 1 \},$$
$$S^1 = \{ (x,y,z) \in \mathbb{R}^3 ; x^2 + y^2 = 1, z = 0 \}.$$

Define $i: S^1 \to C$ as the inclusion and $r: C \to S^1$ by $r(x,y,z) = (x,y,0)$. Obviously $ri = 1: S^1 \to S^1$ whereas $F: C \times I \to C$ defined by $F((x,y,z), t) = (x,y,tz)$ is a homotopy between ir and $1: C \to C$.

The above example leads to some definitions.

13.7   Definition
  A subset A of a topological space X is called a *retract* of X if there is a continuous map $r: X \to A$ such that $ri = 1: A \to A$ (or equivalently if $r|A = 1$), where $i: A \to X$ is the inclusion map. The map r is called a *retraction*.

13.8   Definition
  A subset A of X is called a *deformation retract* of X if there is a retraction $r: X \to A$ such that $ir \simeq 1: X \to X$ where $i: A \to X$ is the inclusion.

In other words A is a deformation retract of X if there is a homotopy F:

$X \times I \to X$ such that $F(x,0) = x$ for all $x \in X$ and $F(x,1) \in A$ for all $x \in X$.

Thus the circle is a deformation retract of the cylinder. Note that if A is a deformation retract of X then A and X are homotopy equivalent. In the example of the circle and the cylinder the map ir is in fact homotopic to the identity relative to the circle. This leads to one more definition.

13.9   Definition

A subset A of X is a *strong deformation retract* if there is a retraction $r: X \to A$ such that $ir \simeq_{rel\ A} 1: X \to X$.

In other words A is a strong deformation retract of X if there is a homotopy $F: X \times I \to X$ such that $F(x,0) = x$ for all $x \in X$, $F(a,t) = a$ for all $a \in A$, $t \in I$ and $F(x,1) \in A$ for all $x \in X$.

A strong deformation retract is, obviously, also a deformation retract. The notion of strong deformation retracts will be useful later on. Warning: some books may call a strong deformation retract simply a deformation retract; we refrain from doing this. Intuitively A is a strong deformation retract of X if X can be deformed, within itself, to A keeping A fixed.

We give one further example involving the notion of strong deformation retract. Consider the subset $Y = C_1 \cup C_2$ of R^2 where

$$C_1 = \{ \ x = (x_1, x_2); (x_1 - 1)^2 + x_2^2 = 1 \ \},$$
$$C_2 = \{ \ x = (x_1, x_2); (x_1 + 1)^2 + x_2^2 = 1 \ \}.$$

Thus Y is a 'figure 8', i.e. a pair of circles joined at one point. Let $X = Y - \{ \ (2,0), (-2,0) \ \}$; then the point $x_0 = (0,0)$ is a strong deformation retract of X. To see this let $i: \{ \ x_0 \ \} \to X$ and $r: X \to \{ \ x_0 \ \}$ denote the obvious maps. Clearly $ri = 1$; to see that $ir \simeq 1$ (rel $\{ \ x_0 \ \}$) we use the following homotopy

$$F: X \times I \to X,$$
$$F(x,s) = (1-s)x / \|((1-s)x_1 + (-1)^i, (1-s)y_2)\| \text{ for } x \in C_i, i = 1,2.$$

Note that $((1-s)x_1 + (-1)^i, (1-s)y_2) \neq (0,0)$ for $x \in X$. It is easy to check that F is continuous. Since $F(x_0,s) = x_0$, $F(x,0) = x$ and $F(x,1) = x_0$ it follows that $ir \simeq 1$ (rel $\{ \ x_0 \ \}$) and so $\{ \ x_0 \ \}$ is a strong deformation retract of X.

13.10   Exercises

    (a) Show that there is a circle in the Möbius strip which is a strong deformation retract of the Möbius strip. Deduce that the Möbius strip and the cylinder are homotopy equivalent.

    (b) Prove that a space X is contractible if and only if the identity map $1: X \to X$ is homotopic to a constant map.

(c) Prove that there is a retraction r: $D^n \to S^{n-1}$ if and only if S^{n-1} is contractible. (Hint: Let F: $S^{n-1} \times I \to S^{n-1}$ be a homotopy between a constant map and the identity map, then use the natural map $S^{n-1} \times I \to D^n$ given by $(x,t) \to tx$ and the fact that $F(S^{n-1} \times \{0\})$ is a single point.)

(d) Prove that if X is connected and has the same homotopy type as Y then Y is also connected.

(e) A subset $A \subseteq X$ is said to be a *weak retract* of X if there exists a continuous map r: $X \to A$ such that ri $\simeq 1$: $A \to A$ where i: $A \to X$ is the inclusion map. A retract is obviously also a weak retract. Give an example of a subset which is a weak retract but is not a retract.

(f) Give an example of a subset that is a deformation retract but is not a strong deformation retract.

(g) A subset $A \subseteq X$ is said to be a *weak deformation retract* of X if the inclusion map i: $A \to X$ is a homotopy equivalence. Thus a deformation retract is also a weak deformation retract. Give an example of a subset that is a weak deformation retract but is not a deformation retract.

(h) Let A be a subspace of X and let Y be a non-empty topological space. Prove that $A \times Y$ is a retract of $X \times Y$ if and only if A is a retract of X.

(i) Prove that the relation 'is a retract of' is transitive (i.e. if A is a retract of B and B is a retract of C, then A is a retract of C).

(j) Prove that the subset $S^1 \times \{x_0\}$ is a retract of $S^1 \times S^1$, but that it is not a strong deformation retract of $S^1 \times S^1$ for any point $x_0 \in S^1$. Is it a deformation retract? Is it a weak deformation retract?

(k) Let $x_0 \in \mathbb{R}^2$. Find a circle in $\mathbb{R}^2$ which is a strong deformation retract of $\mathbb{R}^2 - \{x_0\}$.

(l) Let T be a torus, and let X be the complement of a point in T. Find a subset of X which is homeomorphic to a 'figure 8' curve and which is a strong deformation retract of X.

(m) Prove that S^n is a strong deformation retract of $\mathbb{R}^{n+1} - \{0\}$.

(n) Show that a retract of a Hausdorff space must be a closed subset.

(o) Let Y be a subspace of $\mathbb{R}^n$ and let f,g: $X \to Y$ be two continuous maps. Prove that if for each $x \in X$, $f(x)$ and $g(x)$ can be joined by a straight-line segment in Y then $f \simeq g$. Deduce that any two maps f,g: $X \to \mathbb{R}^n$ must be homotopic.

(p) Let X be any space and let f,g: $X \to S^n$ be two continuous maps such that $f(x) \neq -g(x)$ for all $x \in X$. Prove that $f \simeq g$. (Hint: Consider the map $\mathbb{R}^n - \{ 0 \} \to S^{n-1}$ given by $x \to x/\|x\|$ and use (o) above.) Deduce that if $f: X \to S^n$ is a continuous map that is not surjective then f is homotopic to a constant map.

14

'Multiplication' of paths

If f and g are two paths in X with $f(1) = g(0)$ then by the *product* of f and g we mean the path $f * g$, which is defined as in Chapter 12 by

$$(f * g)(t) = \begin{cases} f(2t) & 0 \leq t \leq \tfrac{1}{2}, \\ g(2t-1) & \tfrac{1}{2} \leq t \leq 1. \end{cases}$$

We shall investigate this 'multiplication' of paths further in this chapter. More precisely we shall look at the multiplication of paths up to homotopy relative to $\{0,1\}$ and see to what extent this multiplication satisfies the axioms for a group.

14.1 Definition

Two paths f, g in X are said to be *equivalent* if f and g are homotopic relative to $\{0,1\}$. We write $f \sim g$.

Note that the paths f_0, f_1 in X are equivalent if there is a continuous map $F: I \times I \to X$ such that

$$F(t,0) = f_0(t) \quad \text{and} \quad F(t,1) = f_1(t) \quad \text{for } t \in I,$$
$$F(0,s) = f_0(0) \quad \text{and} \quad F(1,s) = f_0(1) \quad \text{for } s \in I.$$

See Figure 14.1. In this case we shall write $F: f_0 \sim f_1$.

Figure 14.1

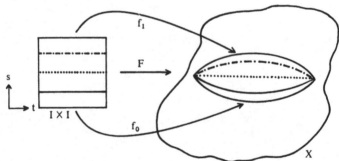

Lemma 13.3 shows that $\sim$ is an equivalence relation on the set of paths in X. We denote the equivalence class of a path f by $[f]$. The first result shows that the product of equivalence classes of paths is well defined by

$$[f]\,[g] = [f * g].$$

14.2   Lemma

Suppose that f_0, f_1, g_0, g_1 are paths in X with $f_0(1) = g_0(0)$ and $f_1(1) = g_1(0)$. If $f_0 \sim f_1$ and $g_0 \sim g_1$ then $f_0 * g_0 \sim f_1 * g_1$.

Proof Let F: $f_0 \sim f_1$ and G: $g_0 \sim g_1$ be the homotopies relative to $\{0,1\}$ realizing the equivalences. Define H: $I \times I \rightarrow X$ by

$$H(t,s) = \begin{cases} F(2t,s) & 0 \le t \le \tfrac{1}{2}, \\[2mm] G(2t-1,s) & \tfrac{1}{2} \le t \le 1, \end{cases}$$

which is continuous by the glueing lemma since $F(1,s) = f_0(1) = g_0(0) = G(0,s)$. It is easy to see that H is a homotopy relative to $\{0,1\}$ between $f_0 * g_0$ and $f_1 * g_1$. See Figure 14.2.

Figure 14.2

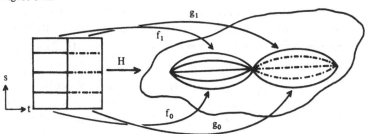

The next result is that multiplication of equivalence classes of paths is associative; in other words

$$([f]\,[g])\,[h] = [f]\,([g]\,[h]).$$

whenever this product makes sense (i.e. if $f(1) = g(0)$ and $g(1) = h(0)$). Note that in general $(f * g) * h \ne f * (g * h)$; see Exercise 14.6(a).

14.3   Lemma

Suppose that f, g, h are three paths in X with $f(1) = g(0)$ and $g(1) = h(0)$. Then $(f * g) * h \sim f * (g * h)$.

Proof Note first that

$$((f * g) * h)(t) = \begin{cases} f(4t) & 0 \leq t \leq \frac{1}{4}, \\ g(4t-1) & \frac{1}{4} \leq t \leq \frac{1}{2}, \\ h(2t-1) & \frac{1}{2} \leq t \leq 1; \end{cases}$$

and

$$(f * (g * h))(t) = \begin{cases} f(2t) & 0 \leq t \leq \frac{1}{2}, \\ g(4t-2) & \frac{1}{2} \leq t \leq \frac{3}{4}, \\ h(4t-3) & \frac{3}{4} \leq t \leq 1. \end{cases}$$

We picture these paths by the diagrams

The diagrams can be used to obtain the algebraic descriptions of the paths in question quite easily. For example consider $(f * g) * h$; when $\frac{1}{4} \leq t \leq \frac{1}{2}$ we use g and compose it with a linear function that changes the interval $[\frac{1}{4}, \frac{1}{2}]$ to $[0,1]$, namely $t \to 4t-1$. In fact any continuous function from $[\frac{1}{4}, \frac{1}{2}]$ to $[0,1]$ which sends $\frac{1}{4}$ to 0 and $\frac{1}{2}$ to 1 will do (see Exercise 14.6(c)) but it is usually easiest to choose a linear function.

To construct a homotopy between $(f * g) * h$ and $f * (g * h)$ consider Figure 14.3. For a given value of s we use f in the interval $[0, (s+1)/4]$, g in the interval $[(s+1)/4, (s+2)/4]$ and h in the interval $[(s+2)/4, 1]$. Using the method described above we are led to defining $F: I \times I \to X$ by

$$F(t,s) = \begin{cases} f((4t)/(1+s)) & 0 \leq t \leq (s+1)/4, \\ g(4t-s-1) & (s+1)/4 \leq t \leq (s+2)/4, \\ h((4t-s-2)/(2-s)) & (s+2)/4 \leq t \leq 1. \end{cases}$$

The function F is continuous and

$$F(t,0) = ((f * g) * h)(t), \qquad F(t,1) = (f * (g * h))(t),$$
$$F(0,s) = f(0) = ((f * g) * h)(0), \qquad F(1,s) = h(1) = ((f * g) * h)(1),$$

so that F provides the required homotopy.

Figure 14.3

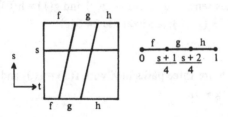

If $x \in X$ then we have defined $\epsilon_x : I \to X$ as the constant path, i.e. $\epsilon_x(t) = x$. The equivalence path of the constant path behaves as a (left or right) identity, i.e.

$$[\epsilon_x][f] = [f] = [f][\epsilon_y]$$

if f is a path that begins at x and ends at y. This is proved in the next result.

14.4   Lemma

If f is a path in X that begins at x and ends at y then $\epsilon_x * f \sim f$ and $f * \epsilon_y \sim f$.

Proof We shall only prove that $\epsilon_x * f \sim f$. The proof that $f * \epsilon_y \sim f$ is quite similar. Consider Figure 14.4. Define F: $I \times I \to X$ by

$$F(t,s) = \begin{cases} x & 0 \leq t \leq (1-s)/2, \\ \\ f((2t-1+s)/(1+s)) & (1-s)/2 \leq t \leq 1. \end{cases}$$

Then $F(t,0) = \epsilon_x * f$, $F(t,1) = f(t)$ and F is a homotopy relative to $\{0,1\}$.

Figure 14.4

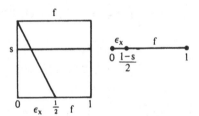

Finally we would like inverses to paths (up to equivalence of paths). To this end recall that if f is a path then $\bar{f}$ is the path defined by $\bar{f}(t) = f(1-t)$. Note that $f \sim g$ if and only if $\bar{f} \sim \bar{g}$ (this is easy to prove). The next result will show that the equivalence class of $\bar{f}$ acts as an inverse for the equivalence class of f, i.e.

$$[f][\bar{f}] = [\epsilon_x] , \quad [\bar{f}][f] = [\epsilon_y]$$

for a path f beginning at x and ending at y.

14.5   Lemma

Let f be a path in X that begins at x and ends at y. Then $f * \bar{f} \sim \epsilon_x$ and $\bar{f} * f \sim \epsilon_y$.

Proof We shall only prove that $f * \bar{f} \sim \epsilon_x$. The path $f * \bar{f}$ is given by

$$(f * \bar{f})(t) = \begin{cases} f(2t) & 0 \leq t \leq \frac{1}{2}, \\ \\ f(2-2t) & \frac{1}{2} \leq t \leq 1. \end{cases}$$

It represents a path in which we travel along f for the first half of our time interval and then in the opposite direction along f for the second half. To make sure that we get from x to y and back to x we travel at speed 2 (i.e. twice 'normal' speed). If we now vary the speed proportionally to (1-s) for $s \in I$ then for each s we get a path that starts at x, goes to $f(2(1-s))$ and then returns to x. For s = 0 we get $f * \bar{f}$ and for s = 1 we get ϵ_x. Define therefore $F: I \times I \to X$ by

$$F(t,s) = \begin{cases} f(2t(1-s)) & 0 \leq t \leq \frac{1}{2}, \\ \\ f((2-2t)(1-s)) & \frac{1}{2} \leq t \leq 1. \end{cases}$$

F is obviously continuous and

$$F(t,0) = (f * \bar{f})(t), \qquad\qquad F(t,1) = f(0) = \epsilon_x(t)$$
$$F(0,s) = f(0) = (f * \bar{f})(0), \qquad F(1,s) = f(0) = (f * \bar{f})(1)$$

so that $f * \bar{f} \sim \epsilon_x$.

An alternative homotopy between $f * \bar{f}$ and ϵ_x is given by $G: I \times I \to X$ where

$$G(t,s) = \begin{cases} f(2t) & 0 \leq t \leq (1-s)/2, \\ f(1-s) & (1-s)/2 \leq t \leq (1+s)/2, \\ f(2-2t) & (1+s)/2 \leq t \leq 1. \end{cases}$$

The idea here is that the time that we spend travelling along f is proportional to (1-s). Thus we go along f for the first (1-s)/2 part of our time interval, then wait at the point f(1-s) and then finally return along f for the last (1-s)/2 portion of our time. Thus when s = 0 this is $f * \bar{f}$ but when s = 1 we spend all our time waiting at x, i.e. ϵ_x.

We shall return to equivalence classes of paths and their products in the next chapter.

14.6   Exercises

(a) Give examples of paths f, g, h in some space X with f(1) = g(0), g(1) = h(0) and (i) with $(f * g) * h \neq f * (g * h)$, (ii) with $(f * g) * h = f * (g * h)$.

(b) Give a direct proof of the result $\epsilon_{f(0)} * f \sim f * \epsilon_{f(1)}$.

(c) Let f be a path in X and let h: $I \to I$ be a continuous mapping with $h(0) = 0, h(1) = 1$. Prove that $f \sim fh$.

(d) Use (c) above to give a direct proof that $f \sim \epsilon_x * f$ where f is a path that begins at x.

(e) Let f, g: $I \to X$ be two paths in X from x to y. Prove that $f \sim g$ if and only if $f * \bar{g} \sim \epsilon_x$.

(f) Suppose that h: $I \to I$ is a continuous function such that $h(0) = 1$ and $h(1) = 0$. Prove that if f is a path then $\bar{f} \sim fh$.

(g) Suppose that $0 = t_0 \leq t_1 \leq t_2 = 1$ and that f: $I \to X$ is a path. Define paths f_1, f_2 by

$f_1(t) = f((1-t)t_0 + tt_1)$,
$f_2(t) = f((1-t)t_1 + tt_2)$.

Prove that $f_1 * f_2 \sim f$. (Hint: Use (c) above.)

(h) Suppose that $0 = t_0 \leq t_1 \leq t_2 \leq ... \leq t_q = 1$ and that f: $I \to X$ is a path. Define paths $f_1, f_2, ..., f_q$ by

$f_i(t) = f((1-t)t_{i-1} + tt_i)$.

Prove that $[f] = [f_1] [f_2] ... [f_q]$.

(i) Suppose that X is a space and that $X = U \cup V$ with U, V open subsets. Show that if f is a path in X then $[f]$ can be expressed as

$[f] = [f_1] [f_2] ... [f_q]$

with each f_i being either a path in U or a path in V. (Hint: Consider the open cover $\{ f^{-1}(U), f^{-1}(V) \}$ of I, rewrite $f^{-1}(U)$ and $f^{-1}(V)$ as disjoint unions of open intervals and use the compactness of I; alternatively use Exercise 7.13(g). Finally use result (h) above.)

(j)   (i) Prove that if h: $(0,1) \to (0,1)$ is a homeomorphism then there exists a homeomorphism f: $[0,1]$ such that $f| (0,1) = h$. Moreover, prove that f is the unique such homeomorphism. (Hint: Look at the closed interval $(0,a]$ (closed in $\overset{\circ}{I}$) and show that $h((0,a])$ is of the form $(0,b]$ or $[c,1)$ for some b or c.)

    (ii) Prove that if h: $I \to I$ is a homeomorphism then h $(\partial I) = \partial I$. (Hint: Use connectedness.)

    (iii) Suppose f, g: $I \to X$ are paths in X so that f: $I \to f(I)$ and g: $I \to g(I)$ are homeomorphisms. Prove that if $f(I) = g(I)$ then either $f \sim g$ or $f \sim \bar{g}$. (Hint: Use (ii) above.)

    (iv) Suppose f, g: $I \to X$ are closed paths in X so that f: $\overset{\circ}{I} \to f(\overset{\circ}{I})$ and g: $\overset{\circ}{I} \to g(\overset{\circ}{I})$ are homeomorphisms. Prove that if $f(I) = g(I)$ and $f(\partial I) = g(\partial I)$ then either $f \sim g$ or $f \sim \bar{g}$.

The fundamental group

From the last chapter we see that the set of equivalence classes of paths (paths are equivalent if they are homotopic relative to { 0,1 }) in a space X appear to satisfy the axioms of a group. The problem is that multiplication is not always defined and the identity 'floats'. The way to get around these problems is to use the concept of a closed path.

15.1   Definition
      A path is said to be *closed* if $f(0) = f(1)$. If $f(0) = f(1) = x$ then we say that f is *based* at x.

Some books use the word 'loop' for a closed path.

Observe that the product $f * g$ is defined for any pair of closed paths based at some point $x \in X$. We denote the set of equivalence classes of closed paths based at $x \in X$ by $\pi(X,x)$. This set has a product defined by $[f] [g] = [f * g]$ for $[f]$, $[g] \in \pi(X,x)$ which is well defined by Lemma 14.2. The next result states that $\pi(X,x)$ is a group; we call it the *fundamental group* of X with base point x.

15.2   Theorem
      $\pi(X,x)$ is a group.

The proof follows from Chapter 14. The product has already been defined. The identity element is $[\epsilon_x]$ (see Lemma 14.4), inverses are given by $[f]^{-1} = [\bar{f}]$ (see Lemma 14.5), while associativity follows from Lemma 14.3.

In view of associativity of the equivalence classes of paths we shall often abbreviate $[(f * g) * h]$ to $[f * g * h]$. Note however that we cannot abbreviate $(f * g) * h$ to $f * g * h$.

Before going any further the reader may like to look at the following exercises.

15.3   Exercises

(a) Why is it not possible to describe $\pi(X,x)$ without reference to the base point?

(b) Show that $\pi(X,x) = 0$ if X is a finite topological space with the discrete topology.

(c) Calculate $\pi(Q,0)$ where Q denotes the set of rationals with the topology induced from the usual topology of $\mathbb{R}$.

(d) Let X be a space for which $\pi(X,x) = 0$. Show that if f, g are two paths in X with $f(0) = g(0) = x$ and $f(1) = g(1)$ then $f \sim g$. (Hint: Use Exercise 14.6(e).)

˙If we choose two different base points x, y $\in$ X then there is no reason, *a priori*, why $\pi(X,x)$ and $\pi(X,y)$ should be related. However, if there is a path from x to y then there is a relation.

15.4   Theorem

Let x, y $\in$ X. If there is a path in X from x to y then the groups $\pi(X,x)$, $\pi(X,y)$ are isomorphic.

Proof Let f be a path from x to y. If g is a closed path based at x then $(\bar{f} * g) * f$ is a closed path based at y. We therefore define

$$u_f: \pi(X,x) \to \pi(X,y)$$

by

$$u_f[g] = [\bar{f} * g * f].$$

This is a homomorphism of groups because

$$
\begin{aligned}
u_f([g][h]) &= u_f[g * h] \\
&= [\bar{f} * g * h * f] \\
&= [\bar{f} * g * f * \bar{f} * h * f] \\
&= [\bar{f} * g * f][\bar{f} * h * f] \\
&= u_f[g] \, u_f[h]
\end{aligned}
$$

Using the path $\bar{f}$ from y to x we can define

$$u_{\bar{f}}: \pi(X,y) \to \pi(X,x)$$

by

$$u_{\bar{f}}[h] = [f * h * \bar{f}].$$

A simple check now shows that $u_{\bar{f}} u_f[g] = [g]$ and $u_f u_{\bar{f}}[h] = [h]$ so that u_f is bijective and hence is an isomorphism.

15.5   Corollary

If X is a path connected space then $\pi(X,x)$ and $\pi(X,y)$ are isomorphic groups, for any pair of points x, y ∈ X.

The above result is not true if we drop the condition that X is path connected; even if X is connected the result does not hold in general. Once we have done a few calculations of fundamental groups (in subsequent chapters) the reader should be able to construct examples of spaces in which $\pi(X,x)$ and $\pi(X,y)$ are not isomorphic for every pair of points x, y ∈ X.

In view of Corollary 15.5 it is tempting to drop the x from $\pi(X,x)$ when X is path connected. This is dangerous as there is no canonical isomorphism from $\pi(X,x)$ to $\pi(X,y)$ since different paths from x to y may give different isomorphisms.

15.6   Exercises

(a) Prove that two paths f, g from x to y give rise to the same isomorphism from $\pi(X,x)$ to $\pi(X,y)$ (i.e. $u_f = u_g$) if and only if $[g * \bar{f}]$ belongs to the centre of $\pi(X,x)$. The *centre* Z(G) of a group G is defined by

$$Z(G) = \{\, a \in G;\, ab = ba \text{ for all } b \in G \,\}.$$

(b) Let $u_f\colon \pi(X,x) \to \pi(X,y)$ be the isomorphism determined by a path f from x to y. Prove that u_f is independent of f if and only if $\pi(X,x)$ is abelian.

For the rest of this chapter we shall be concerned with the effect that a continuous map between topological spaces has upon fundamental groups. Let $\varphi\colon X \to Y$ be a continuous map; the following three facts are obvious.

(i) If f, g are paths in X then φf, φg are paths in Y.
(ii) If $f \sim g$ then $\varphi f \sim \varphi g$.
(iii) If f is a closed path in X based at x ∈ X then φf is a closed path in Y based at $\varphi(x)$.

Thus if $[f] \in \pi(X,x)$ then $[\varphi f]$ is a well-defined element of $\pi(Y,\varphi(x))$. We therefore define

$$\varphi_*\colon \pi(X,x) \to \pi(Y,\varphi(x))$$

by

$$\varphi_*\,[f] = [\varphi f].$$

15.7   Lemma

φ_* is a homomorphism of groups.

The proof is easy: $\varphi_*\,([f]\,[g]) = \varphi_*\,[f * g] = [\varphi(f * g)] = [\varphi f * \varphi g] = [\varphi f]\,[\varphi g] = \varphi_*\,[f]\,\varphi_*\,[g].$

15.8   **Definition**

The homomorphism φ_*: $\pi(X,x) \to \pi(Y,\varphi(x))$ defined by φ_* [f] = [φf], where φ: X → Y is a continuous map, is called the *induced* homomorphism.

The next two results are easy to prove and are left for the reader.

15.9   **Theorem**

(i) Suppose φ: X → Y and ψ: Y → Z are continuous maps, then $(\psi\varphi)_* = \psi_*\varphi_*$.

(ii) If 1: X → X is the identity map then 1_* is the identity homomorphism on $\pi(X,x)$.

15.10   **Corollary**

If φ: X → Y is a homeomorphism then φ_*: $\pi(X,x) \to \pi(Y,\varphi(x))$ is an isomorphism.

Thus the fundamental group provides a means of going from topology to algebra. This process has the following features.

    (i) For each topological space (with some base point) we get a group (the fundamental group).

    (ii) For each continuous map between topological spaces we get a homomorphism (the induced homomorphism) between groups.

    (iii) The composite of continuous maps induces the composite of the induced homomorphisms.

    (iv) The identity map induces the identity homomorphism.

    (v) A homeomorphism induces an isomorphism.

This provides a good example of what *algebraic topology* is about. We replace topology by algebra and then use our knowledge of algebra to learn something about topology. Of course if the fundamental groups of two spaces are isomorphic it does not mean that the spaces are homeomorphic. However, if the fundamental groups are not isomorphic then the spaces cannot be homeomorphic.

Remark: The features (i) − (v) mentioned above are an example of a *functor*. Thus the fundamental group is a functor from topology (especially the collection of topological spaces with base points and base point preserving continuous maps) to algebra (especially the collection of groups and group homomorphisms).

15.11   Exercises

(a) Give an example of an injective continuous map $\varphi: X \to Y$ for which φ_* is not injective. (Assume that $\pi(S^1, x) \cong \mathbb{Z}$, $\pi(D^2, x) = 0$.)

(b) Give an example of a surjective continuous map $\varphi: X \to Y$ for which φ_* is not surjective.

(c) Prove that if $\varphi: X \to Y$ is continuous and f is a path from x to y then $\varphi_* \, u_f = u_{\varphi(f)} \, \varphi: \pi(X, x) \to \pi(Y, \varphi(y))$, where u_f and $u_{\varphi(f)}$ are the isomorphisms of fundamental groups determined by f and $\varphi(f)$.

(d) Prove that two continuous mappings $\varphi, \psi: X \to Y$, with $\varphi(x_0) = \psi(x_0)$ for some point $x_0 \in X$, induce the same homomorphism from $\pi(X, x_0)$ to $\pi(Y, \varphi(x_0))$ if φ and ψ are homotopic relative to x_0.

(e) Suppose that A is a retract of X with retraction r: $X \to A$. Prove that

$$i_*: \pi(A, a) \to \pi(X, a)$$

is a monomorphism (where i: $A \to X$ denotes inclusion) and that

$$r_*: \pi(X, a) \to \pi(A, a)$$

is an epimorphism for any point $a \in A$.

(f) With the notation of (e) above, suppose that $i_* \, \pi(A, a)$ is a normal subgroup of $\pi(X, a)$. Prove that $\pi(X, a)$ is the direct product of the subgroups image (i_*) and kernel (r_*).

(g) Prove that if A is a strong deformation retract of X then the inclusion map i: $A \to X$ induces an isomorphism

$$i_*: \pi(A, a) \to \pi(X, a)$$

for any point $a \in A$.

(h) Show that if $\varphi: X \to X$ is a continuous map with $\varphi \simeq 1$ then

$$\varphi_*: \pi(X, x_0) \to \pi(X, \varphi(x_0))$$

is an isomorphism for each point $x_0 \in X$. (In desperation look at the proof of Theorem 15.12.)

The next result generalizes Exercise 15.11(d).

15.12   Theorem

Let $\varphi, \psi: X \to Y$ be continuous mappings between topological spaces and let F: $\varphi \simeq \psi$ be a homotopy. If f: $I \to Y$ is the path from $\varphi(x_0)$ to $\psi(x_0)$ given by $f(t) = F(x_0, t)$ then the homomorphisms

$$\varphi_*: \pi(X, x_0) \to \pi(Y, \varphi(x_0))$$

and

$$\psi_*: \pi(X, x_0) \to \pi(Y, \psi(x_0))$$

are related by $\psi_* = u_f\, \varphi_*$ where u_f is the isomorphism from $\pi(Y,\varphi(x_0))$ to $\pi(Y,\psi(x_0))$ determined by the path f.

Proof We have to show that if $[g] \in \pi(X,x_0)$ then $[\psi g] = [\bar{f} * \varphi g * f]$. In other words we have to show that the paths $(\bar{f} * \varphi g) * f$ and ψg are equivalent. Observe that

$$((\bar{f} * \varphi g) * f)(t) = \begin{cases} f(1-4t) & 0 \le t \le \frac{1}{4}, \\ \varphi g(4t-1) & \frac{1}{4} \le t \le \frac{1}{2}, \\ f(2t-1) & \frac{1}{2} \le t \le 1, \end{cases}$$

which we may rewrite as

$$((\bar{f} * \varphi g) * f)(t) = \begin{cases} F(x_0, 1-4t) & 0 \le t \le \frac{1}{4}, \\ F(g(4t-1),0) & \frac{1}{4} \le t \le \frac{1}{2}, \\ F(x_0, 2t-1) & \frac{1}{2} \le t \le 1. \end{cases}$$

Meanwhile

$$\psi g(t) = F(g(t),1)$$

The way to see a homotopy between $(\bar{f} * \varphi g) * f$ and ψg is to note that the path ψg is equivalent to $(\epsilon_x * \psi g) * \epsilon_x$ where $x = \psi(x_0)$. The path $(\epsilon_x * \psi g) * \epsilon_x$ has the form

$$((\epsilon_x * \psi g) * \epsilon_x)(t) = \begin{cases} F(x_0,1) & 0 \le t \le \frac{1}{4}, \\ F(g(4t-1),1) & \frac{1}{4} \le t \le \frac{1}{2}, \\ F(x_0,1) & \frac{1}{2} \le t \le 1. \end{cases}$$

Define therefore a map $H: I \times I \to Y$ by

$$H(t,s) = \begin{cases} F(x_0, 1-4t(1-s)) & 0 \le t \le \frac{1}{4}, \\ F(g(4t-1),s) & \frac{1}{4} \le t \le \frac{1}{2}, \\ F(x_0, 1 + 2(t-1)(1-s)) & \frac{1}{2} \le t \le 1. \end{cases}$$

The map H is clearly continuous with

$$\begin{aligned} H(t,0) &= ((\bar{f} * \varphi g) * f)(t), \\ H(t,1) &= ((\epsilon_x * \psi g) * \epsilon_x)(t), \\ H(0,s) &= F(x_0,1) = \psi(x_0), \\ H(1,s) &= F(x_0,1) = \psi(x_0). \end{aligned}$$

Hence $(\bar{f} * \varphi g) * f \sim (\epsilon_x * \psi g) * \epsilon_x \sim \psi g$ which proves that $u_f\, \varphi_* = \psi_*$.

Another way of stating the above result is that there is a commutative diagram

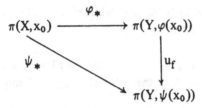

The next two results concern homotopy equivalent spaces.

15.13   Theorem

If $\varphi: X \to Y$ is a homotopy equivalence then $\varphi_*: \pi(X,x) \to \pi(Y,\varphi(x))$ is an isomorphism for any $x \in X$.

Proof Since φ is a homotopy equivalence there is a continuous map $\psi: Y \to X$ such that $\varphi\psi \simeq 1: Y \to Y$ and $\psi\varphi \simeq 1: X \to X$. By Theorem 15.12 we have $u_f (\psi\varphi)_* = 1_*$, and since u_f and 1_* are isomorphisms so is $(\psi\varphi)_* = \psi_* \varphi_*$. This means that ψ_* is an epimorphism and φ_* is a monomorphism. Similarly $\varphi_* \psi_*$ is an isomorphism, which means that φ_* is an epimorphism and ψ_* is a monomorphism. This proves the result.

15.14   Corollary

A contractible space has trivial fundamental group.

We have a special name for path connected spaces with trivial fundamental group.

15.15   Definition

A topological space X is *simply connected* if it is path connected and $\pi(X,x) = \{ 1 \}$ for some (and hence any) $x \in X$.

Thus contractible spaces are simply connected. The converse is not true as the reader will discover later on.

15.16   Exercises

(a) Suppose that A is a weak retract of X (see Exercise 13.10(e)). What can you say about the following homomorphisms

$i_*: \pi(A,a) \to \pi(X,a)$,

$r_*: \pi(X,a) \to \pi(A,a)$,

for $a \in A$?

(b) A space X is said to have property C if for every closed path f: $I \to X$ there is a homotopy $F: I \times I \to X$ such that

$F(t,0) = f(t)$, $F(t,1)$ is constant,
$F(0,s) = F(1,s)$ for all $s \in I$.

Note that F is not necessarily a homotopy relative to $\{0,1\}$. Prove that if X has property C then X is simply connected.

(c) Suppose that $X = U \cup V$ with U,V open and simply connected and $U \cap V$ path connected. Prove that X is simply connected. Hence prove that if $n \geq 2$ then S^n is simply connected. (Hint: Use Exercises 14.6(i) and (e).)

The last result that we prove in this chapter concerns the fundamental group of the topological product of two spaces. It could have been included much earlier in the chapter.

15.17   Theorem

Let X, Y be two path connected topological spaces. The fundamental group of the product $X \times Y$ is isomorphic to the product of the fundamental groups of X and Y.

Proof Let p: $X \times Y \to X$, q: $X \times Y \to Y$ denote the projection maps. Define

$$\varphi: \pi(X \times Y, (x_0, y_0)) \to \pi(X, x_0) \times \pi(Y, y_0)$$

by

$$\varphi[f] = (p_*[f], q_*[f]) = ([pf], [qf]).$$

First we check that φ is well defined. If $f \sim g$ then there is a continuous map F: $I \times I \to X \times Y$ such that $F(t,0) = f(t)$, $F(t,1) = g(t)$ and $F(0,s) = F(1,s) = (x_0,y_0)$. The continuous maps pF: $I \times I \to X$ and qF: $I \times I \to Y$ provide the equivalences pf $\sim$ pg and qf $\sim$ qg so that $\varphi[f] = \varphi[g]$ and φ is well defined.

To see that φ is surjective suppose that $([f_1], [f_2])$ belongs to $\pi(X,x_0) \times \pi(Y,y_0)$. Consider f: $I \to X \times Y$ given by $f(t) = (f_1(t), f_2(t))$. Clearly $\varphi[f] = ([f_1], [f_2])$.

To show that φ is injective suppose that $\varphi[f] = \varphi[g]$. This means that pf $\sim$ pg and qf $\sim$ qg. If F_1: $I \times I \to X$ and F_2: $I \times I \to Y$ give these equivalences then F: $I \times I \to X \times Y$, given by $F(t,s) = (F_1(t,s), F_2(t,s))$ provides the equivalence $f \sim g$.

Finally, that φ is a homomorphism follows readily from the obvious fact that if f, g: $I \to X \times Y$ are paths with $f(1) = g(0)$ then $p(f * g) = pf * pg$ and $q(f * g) = qf * qg$.

Alternative ways of proving Theorem 15.17 appear in the exercises.

15.18   Exercises

(a) Prove that the product of two simply connected spaces is simply connected.

(b) Let $f: I \to X$, $g: I \to Y$ be closed paths based at $x_0 \in X$ and $y_0 \in Y$ respectively. Let $i: X \to X \times Y$ and $j: Y \to X \times Y$ be inclusions defined by $i(x) = (x, y_0)$ and $j(y) = (x_0, y)$. Show that the two paths $(if) * (jg)$ and $(jg) * (if)$ in $X \times Y$ are equivalent.

(c) In the notation of (b) above show that the mapping of $\pi(X, x_0) \times \pi(Y, y_0)$ to $\pi(X \times Y, (x_0, y_0))$ given by $([f], [g]) \to [(if) * (jg)]$ is an isomorphism of groups.

(d) A *topological group* G is a group that is also a topological space in which the maps

$$\mu: G \times G \to G \qquad \nu: G \to G$$

defined by $\mu(g_1, g_2) = g_1 g_2$ and $\nu(g) = g^{-1}$ are continuous. Let f, h be closed paths in G based at the identity element e of G. Define $f \cdot h$ by

$$(f \cdot h)(t) = \mu(f(t), h(t)) \qquad t \in I.$$

Prove that

$$f * h \sim f \cdot h \sim h * f$$

and deduce that the fundamental group $\pi(G, e)$ is abelian. Show, furthermore, that the homomorphism

$$\nu_*: \pi(G, e) \to \pi(G, e)$$

satisfies $\nu_* [f] = [f]^{-1}$.

(e) For $S^1 \subseteq \mathbb{C}$ define $\mu: S^1 \times S^1 \to S^1$ by $\mu(z_1, z_2) = z_1 z_2$, and $\nu: S^1 \to S^1$ by $\nu(z) = z^{-1}$. Prove that S^1 is a topological group. Deduce that $\pi(S^1, 1)$ is an abelian group.

(f) This question is a generalization of (d) above. Let x_0 be a point of the space X. Suppose that there is a continuous map $\mu: X \times X \to X$ such that

$$\mu(x, x_0) = \mu(x_0, x) = x$$

for all $x \in X$. Prove that if $f, g: I \to X$ are closed paths based at $x_0 \in X$ and $i, j: X \to X \times X$ denote the inclusions $i(x) = (x, x_0)$, $j(x) = (x_0, x)$ then

$$\mu((if) * (jg)) = f * g.$$

Deduce, from (b) above, that $\pi(X, x_0)$ is an abelian group.

(g) This question is a generalization of (f) above. A space X is called an *H-space* (after Heinz Hopf) if there is a continuous map $\mu: X \times X \to X$ and a point $x_0 \in X$ such that $\mu i \simeq 1$ (rel x_0) and $\mu j \simeq 1$ (rel x_0)

where i and j are the inclusion maps as in (f). Note that $\mu(x_0,x_0) = x_0$. Prove that the fundamental group $\pi(X,x_0)$ of an H-space is abelian. (Hint: Show that $\mu((if) * (jg)) \sim f * g$.)

The fundamental group $\pi(X,x_0)$ is often denoted by $\pi_1(X,x_0)$, the '1' coming from the fact that we used paths (maps from $I \subseteq \mathbb{R}^1$) to define the fundamental group. More generally we can define $\pi_n(X,x_0)$ by using maps from $I^n \subseteq \mathbb{R}^n$ to X. This is called the *n-th homotopy group* of X at x_0. We can now briefly indicate the appropriate definitions; the uninterested reader can go straight to the next chapter.

Let ∂I^n denote, as usual, the boundary of I^n, i.e.

$$\partial I^n = \{ (t_1, t_2, ..., t_n) \in I^n ; t_i = 0 \text{ or } 1 \text{ for some } i \}.$$

The set $\pi_n(X,x_0)$ consists of homotopy classes relative to ∂I^n of a continuous map f: $I^n \to X$ with $f(\partial I^n) = x_0$. A product is defined by

$$[f] [g] = [f * g]$$

where

$$(f * g)(t) = \begin{cases} f(2t_1, t_2, ... t_n) & 0 \leq t_1 \leq \tfrac{1}{2}, \\ g(2t_1 - 1, t_2, ..., t_n) & \tfrac{1}{2} \leq t_1 \leq 1. \end{cases}$$

It can be easily verified that the product is well defined and that it gives $\pi_n(X,x_0)$ the structure of a group. Of course if n = 1 then we get the fundamental group. The fundamental group is not necessarily an abelian group as we shall see later; however, $\pi_n(X,x_0)$ is always an abelian group if $n \geq 2$.

15.19 Exercises

(a) Prove that $\pi_n(X,x_0)$ is a group.

(b) Prove that if there is a path in X from x_0 to x_1 then $\pi_n(X,x_0)$ and $\pi_n(X,x_1)$ are isomorphic.

Figure 15.1

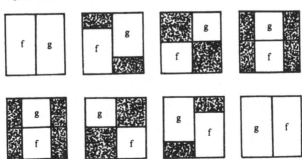

(c) For a continuous map $\varphi: X \to Y$ define $\varphi_*: \pi_n(X, x_0) \to \pi_n(Y, \varphi(x_0))$ and prove that φ_* is a homomorphism. Prove also Theorem 15.9 for the n-th homotopy groups. Deduce that homeomorphic spaces have isomorphic homotopy groups.

(d) Prove that homotopy equivalent spaces have isomorphic homotopy groups.

(e) Prove that if $n \geq 2$ then $\pi_n(X, x_0)$ is an abelian group. (Hint: A homotopy between $f * g$ and $g * f$ is suggested in Figure 15.1.)

Remark: There is a sort of converse to (d) above. This is a theorem (due to J.H.C. Whitehead) which states that if X and Y are a certain type of topological space (the so-called path connected CW complexes) and if $\varphi: X \to Y$ is a continuous map that induces isomorphisms $\varphi_*: \pi_n(X, x_0) \to \pi_n(Y, \varphi(x_0))$ for all $n \geq 1$ then φ is a homotopy equivalence.

The fundamental group of a circle

Except for some trivial cases we have not, so far, calculated the fundamental group of a space. In this chapter we shall calculate the fundamental group of the circle S^1, the answer being $\mathbb{Z}$ the integers. Intuitively we see this result as follows. A closed path f in S^1 based at $1 \in S^1$ winds a certain number of times around the circle; this number is called the winding number or degree of f. (Start with $f(0) = 1$ and consider $f(t)$ as t increases; every time we go once around the circle in an anticlockwise direction record a score of +1, every time we go once around in a clockwise direction score −1. The total score is the winding number or degree of f.) Thus to each closed path f based at 1 we get an integer. It turns out that two closed paths are equivalent (i.e. homotopic rel { 0,1 }) if and only if their degrees agree. Finally, for each integer n there is a closed path of degree n.

To get a more precise definition of the degree of a closed path we consider the real numbers $\mathbb{R}$ mapping onto S^1 as follows.

$$e: \mathbb{R} \to S^1,$$
$$t \to \exp(2\pi i t).$$

Figure 16.1

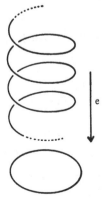

e

Geometrically we think of the reals as a spiral with e being the projection mapping (see Figure 16.1). Note that $e^{-1}(1) = \mathbb{Z} \subseteq \mathbb{R}$. The idea now is that if we are given $f: I \to S^1$ with $f(0) = f(1) = 1$ then we show that there is a unique map $\tilde{f}: I \to \mathbb{R}$ with $\tilde{f}(0) = 0$ and $e\tilde{f} = f$ (the map $\tilde{f}$ is called a *lift* of f) Since $f(1) = 1$ we must have $\tilde{f}(1) \in e^{-1}(1) = \mathbb{Z}$; this integer is defined to be the degree of f. We then go on to show that if f_0 and f_1 are equivalent paths in S^1 then $\tilde{f}_0(1) = \tilde{f}_1(1)$. This leads to a function $\pi(S^1,1) \to \mathbb{Z}$ which we finally show is an isomorphism of groups.

The 'method of calculation' of $\pi(S^1,1)$ that we shall be presenting generalizes to some other spaces; see the subsequent three chapters. In fact the next lemma is the starting point for a crucial definition in Chapter 17.

16.1   Lemma

Let U be any open subset of $S^1 - \{ 1 \}$ and let $V = I \cap e^{-1}(U) \subseteq \mathbb{R}$. Then $e^{-1}(U)$ is the disjoint union of the open sets $V + n = \{ v+n; v \in V \}$, $n \in \mathbb{Z}$, each of which is mapped homeomorphically onto U by e.

Proof We assume that U is an open interval, i.e.

$$U = \{ \exp(2\pi it); 0 \leq a < t < b \leq 1 \}$$

for some a,b. Then $V = (a,b)$ and $V + n = (a+n, b+n)$. It is clear that $e^{-1}(U)$ is the disjoint union of the open sets $V + n$ ($n \in \mathbb{Z}$). Let e_n denote the restriction of e to $(a+n, b+n)$. Clearly e_n is continuous and bijective. To check that e_n^{-1} is continuous we consider $(a+n, b+n)$ and let $W \subseteq (a+n, b+n)$ be a closed (and hence compact) subset. Since W is compact and S^1 is Hausdorff, e_n induces a homeomorphism $W \to e_n(W)$ by Theorem 8.8. In particular $e_n(W)$ is compact and hence closed. This shows that if W is a closed subset then $e_n(W)$ is also closed; thus e_n^{-1} is continuous and hence e_n is a homeomorphism.

16.2   Exercise

Show that the above holds for $S^1 - \{ x \}$, where x is any point of S^1.

16.3   Corollary

If $f: X \to S^1$ is not surjective then f is null homotopic.

Proof If $x \notin$ image (f) then $S^1 - \{ x \}$ is homeomorphic to $(0,1)$ which is contractible. ($x = \exp(2\pi is)$ for some s and $S^1 = \{ \exp(2\pi it); s \leq t < 1+s \}$.)

We come now to the first major result of this chapter: the so-called *path lifting theorem* (for e: $\mathbb{R} \to S^1$).

16.4   Theorem

Any continuous map $f: I \to S^1$ has a lift $\widetilde{f}: I \to \mathbb{R}$. Furthermore given $x_0 \in \mathbb{R}$ with $e(x_0) = f(0)$ there is a unique lift $\widetilde{f}$ with $\widetilde{f}(0) = x_0$.

Proof   For each $x \in S^1$ let U_x be an open neighbourhood of x such that $e^{-1}(U_x)$ is the disjoint union of open subsets of $\mathbb{R}$ each of which are mapped homeomorphically onto U_x by e. The set $\{ f^{-1}(U_x); x \in S^1 \}$ may be expressed in the form $\{ (x_j, y_j) \cap I; j \in J \}$ which is an open cover of I. Since I is compact there is a finite subcover of the form

$$[0, t_1 + \epsilon_1), (t_2 - \epsilon_2, t_2 + \epsilon_2),..., (t_n - \epsilon_n, 1]$$

with $t_i + \epsilon_i > t_{i+1} - \epsilon_{i+1}$ for i = 1,2,...,n - 1. Now choose $a_i \in (t_{i+1} - \epsilon_{i+1}, t_i + \epsilon_i)$ for i = 1,2,...,n - 1 so that

$$0 = a_0 < a_1 < a_2 < ... < a_n = 1.$$

Obviously $f([a_i, a_{i+1}]) \subset S^1$, but more so $f([a_i, a_{i+1}])$ is contained in an open subset S_i of S^1 such that $e^{-1}(S_i)$ is the disjoint union of open subsets of $\mathbb{R}$ each of which are mapped homeomorphically onto S_i by e.

We shall define liftings $\widetilde{f}_k$ inductively over $[0, a_k]$ for k = 0,1,...,n such that $\widetilde{f}_k(0) = x_0$. For k = 0 this is trivial: $\widetilde{f}_0(0) = x_0$; we have no choice.

Figure 16.2

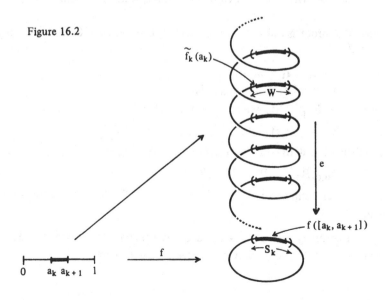

Suppose that $\tilde{f}_k$: $[0,a_k] \rightarrow \mathbb{R}$ is defined and is unique. Recall that $f([a_k,a_{k+1}]) \subseteq S_k$ and that $e^{-1}(S_k)$ is the disjoint union of $\{ W_j; j \in J \}$ with $e|W_j$: $W_j \rightarrow S_k$ being a homeomorphism for each $j \in J$. Now $\tilde{f}_k(a_k) \in W$ for some unique member W of $\{ W_j; j \in J \}$; see Figure 16.2. Any extension $\tilde{f}_{k+1}$ must map $[a_k,a_{k+1}]$ into W since $[a_k,a_{k+1}]$ is path connected. Since the restriction $e|W$: $W \rightarrow S_k$ is a homeomorphism there is a unique map ρ: $[a_k,a_{k+1}] \rightarrow W$ such that $e\rho = f|[a_k,a_{k+1}]$ (in fact $\rho = (e|W)^{-1} f$). Now define $\tilde{f}_{k+1}$ by

$$\tilde{f}_{k+1}(s) = \begin{cases} \tilde{f}_k(s) & 0 \le s \le a_k, \\ \\ \rho(s) & a_k \le s \le a_{k+1}, \end{cases}$$

which is continuous by the glueing lemma since $\tilde{f}_k(a_k) = \rho(a_k)$ and is unique by construction. By induction we obtain $\tilde{f}$.

Using this theorem we can define the degree of a closed path in S^1. Let f be a closed path in S^1 based at 1 and let $\tilde{f}$: $I \rightarrow \mathbb{R}$ be the unique lift with $\tilde{f}(0) = 0$. Since $e^{-1}(f(1)) = e^{-1}(1) = \mathbb{Z}$ we see that $\tilde{f}(1)$ is an integer which is defined to be the *degree* of f. To show that equivalent paths have the same degree we shall first show that equivalent paths have equivalent lifts. To do this we replace I by I^2 in the previous theorem to obtain.

16.5   Lemma
Any continuous map F: $I^2 \rightarrow S^1$ has a lift $\tilde{F}$: $I^2 \rightarrow \mathbb{R}$. Furthermore given $x_0 \in \mathbb{R}$ with $e(x_0) = F(0,0)$ there is a unique lift $\tilde{F}$ with $\tilde{F}(0,0) = x_0$.

Proof The proof is quite similar to that of Theorem 16.4. Since I^2 is compact we find

$$0 = a_0 < a_1 < ... < a_n = 1,$$
$$0 = b_0 < b_1 < ... < b_m = 1,$$

such that $F(R_{i,j}) \subseteq S^1$, where $R_{i,j}$ is the rectangle

$$R_{i,j} = \{ (t,s) \in I^2 ; a_i \le t \le a_{i+1}, b_j \le s \le b_{j+1} \}.$$

The lifting $\tilde{F}$ is defined inductively over the rectangles

$$R_{0,0}, R_{0,1}, ..., R_{0,m}, R_{1,0}, R_{1,1}, ...$$

by a process similar to that in Theorem 16.4. We leave the details for the reader.

As a corollary we have the so-called *monodromy theorem* for e: $\mathbb{R} \rightarrow S^1$, which tells us that equivalent paths have the same degree.

16.6   Corollary
Suppose that f_0 and f_1 are equivalent paths in S^1 based at 1. If $\tilde{f}_0$ and $\tilde{f}_1$ are lifts with $\tilde{f}_0(0) = \tilde{f}_1(0)$ then $\tilde{f}_0(1) = \tilde{f}_1(1)$.

Proof Let F be the homotopy rel $\{\,0,1\,\}$ between f_0 and f_1. It lifts uniquely to $\tilde{F}: I^2 \rightarrow \mathbb{R}$ with $\tilde{F}(0,0) = \tilde{f}_0(0) = \tilde{f}_1(0)$. Since $F(t,0) = f_0(t)$ and $F(t,1) = f_1(t)$, we have $\tilde{F}(t,0) = \tilde{f}_0(t)$ and $\tilde{F}(t,1) = \tilde{f}_1(t)$. Also, $\tilde{F}(1,t)$ is a path from $\tilde{f}_0(1)$ to $\tilde{f}_1(1)$ since $F(1,t) = f_0(1) = f_1(1)$. But $\tilde{F}(1,t) \in e^{-1}$ $(f_0(1)) \cong \mathbb{Z}$, which means that $\tilde{F}(1,t)$ is constant and hence $\tilde{f}_0(1) = \tilde{f}_1(1)$ thus completing the proof. Note that in fact $\tilde{F}$ provides a homotopy rel $\{\,0,1\,\}$ between $\tilde{f}_0$ and $\tilde{f}_1$.

We are now in a position to calculate the fundamental group of the circle.

16.7   Theorem
$$\pi(S^1,1) \cong \mathbb{Z}.$$

Proof Define $\varphi: \pi(S^1,1) \rightarrow \mathbb{Z}$ by $\varphi([f]) = \deg(f)$, the degree of f. Recall that $\deg(f) = \tilde{f}(1)$ where $\tilde{f}$ is the unique lift of f with $\tilde{f}(0) = 0$. The function φ is well defined by Corollary 16.6. We shall show that φ is an isomorphism of groups.

First we show that φ is a homomorphism. Let $\ell_a(f)$ denote the lift of f beginning at $a \in e^{-1}(f(0))$. Thus $\ell_0(f) = \tilde{f}$ and $\ell_a(f)(t) = \tilde{f}(t) + a$ for a path in S^1 beginning at 1. It is clear that

$$\ell_a(f * g) = \ell_a(f) * \ell_b(g)$$

where $b = \tilde{f}(1) + a$. Thus if $[f], [g] \in \pi(S^1,1)$ then

$$
\begin{aligned}
\varphi([f]\,[g]) &= \varphi([f * g] = \widetilde{f * g}(1) \\
&= \ell_0(f * g)(1) \\
&= (\ell_0(f) * \ell_b(g))(1) \text{ where } b = \tilde{f}(1) \\
&= \ell_b(g)(1) \\
&= b + \tilde{g}(1) \\
&= \tilde{f}(1) + \tilde{g}(1) \\
&= \varphi([f]) + \varphi([g])
\end{aligned}
$$

which shows that φ is a homomorphism.

To show that φ is surjective is rather easy: given $n \in \mathbb{Z}$ let $g: I \rightarrow \mathbb{R}$ be given by $g(t) = nt$; then $eg: I \rightarrow S^1$ is a closed path based at 1. Since g is the lift of eg with $g(0) = 0$ we have $\varphi([eg]) = \deg(eg) = g(1) = n$ which shows that φ is surjective.

To show that φ is injective we suppose that $\varphi([f]) = 0$, i.e. $\deg(f) = 0$.

This means that the lift $\tilde{f}$ of f satisfies $\tilde{f}(0) = \tilde{f}(1) = 0$. Since $\mathbb{R}$ is contractible we have $\tilde{f} \simeq \epsilon_0$ (rel $\{ 0,1 \}$); in other words there is a map $F: I^2 \to \mathbb{R}$ with $F(0,t) = \tilde{f}(t)$, $F(1,t) = 0$ and $F(t,0) = F(t,1) = 0$. Indeed $F(s,t) = (1-s) \tilde{f}(t)$. But $eF: I^2 \to S^1$ with $eF(0,t) = f(t)$, $eF(1,t) = 1$, $eF(t,0) = eF(t,1) = 1$ and so $f \simeq \epsilon_1$ (rel $\{ 0,1 \}$), i.e. $[f] = 1 \in \pi(S^1,1)$, which proves that φ is injective and hence φ is an isomorphism.

This completes the proof of the main result of this chapter. As a corollary we immediately obtain:

16.8   Corollary
The fundamental group of the torus is $\mathbb{Z} \times \mathbb{Z}$.

We close the chapter by giving two applications. The first is well known and is the *fundamental theorem of algebra*.

16.9   Corollary
Every non-constant complex polynomial has a root.

Proof We may assume without loss of generality that our polynomial has the form

$$p(z) = a_0 + a_1 z + ... + a_{k-1} z^{k-1} + z^k$$

with $k \geq 1$. Assume that p has no zero (i.e. no root). Define a function G: $I \times [0,\infty) \to S^1 \subset \mathbb{C}$ by

$$G(t,r) \;=\; \frac{p(r \exp(2\pi it))}{|p(r \exp(2\pi it))|} \; \frac{|p(r)|}{p(r)}$$

for $0 \leq t \leq 1$ and $r \geq 0$. Clearly G is continuous. Define $F: I^2 \to S^1$ by

$$F(t,s) = \begin{cases} G(t,s/(1-s)) & 0 \leq t \leq 1, 0 \leq s < 1, \\[2mm] \exp(2\pi ikt) & 0 \leq t \leq 1, s = 1. \end{cases}$$

By observing that

$$\lim_{s \to 1} F(t,s) = \lim_{s \to 1} G(t,s/(1-s)) = \lim_{r \to \infty} G(t,r) = (\exp(2\pi it))^k$$

we see that F is continuous. Also, we see that F is a homotopy rel $\{ 0,1 \}$ between $f_0(t) = F(t,0)$ and $f_1(t) = F(t,1)$. But $f_0(t) = 1$ and $f_1(t) = \exp (2\pi ikt)$, so that $\deg(f_0) = 0$ while $\deg(f_1) = k$, which is a contradiction (unless $k = 0$).

The second application comes under the title of *Brouwer's fixed point theorem in the plane*. Recall that in Chapter 10 we proved a fixed point theorem for I; the next result is the analogous theorem for D^2. The result is also true in higher dimensions but the proof requires tools other than the fundamental group.

16.10 Corollary

Any continuous map f: $D^2 \to D^2$ has a fixed point, i.e. a point x such that f(x) = x.

Figure 16.3

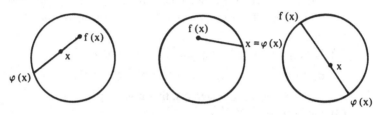

Proof Suppose to the contrary that x ≠ f(x) for all x ∈ D^2. Then we may define a function φ: $D^2 \to S^1$ by setting $\varphi(x)$ to be the point on S^1 obtained from the intersection of the line segment from f(x) to x extended to meet S^1; see Figure 16.3. That φ is continuous is obvious. Let i: $S^1 \to D^2$ denote the inclusion, then $\varphi i = 1$ and we have a commutative diagram

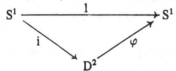

This leads to another commutative diagram

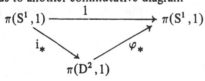

But $\pi(D^2, 1) = 0$, since D^2 is contractible, and so we get a commutative diagram

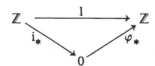

which is impossible. This contradiction proves the result.

16.11   Exercises

(a) Given $[f] \in \pi(S^1, 1)$, let γ be the contour $\{ f(t); t \in I \} \subset \mathbb{C}$ and
define

$$w(f) = \frac{1}{2\pi i} \int_\gamma \frac{dz}{z}$$

Prove that (i) $w(f)$ is an integer,
(ii) $w(f)$ is independent of the choice of $f \in [f]$,
(iii) $w(f) = \deg(f)$.

(b) Let $f: S^1 \to S^1$ be the mapping defined by $f(z) = z^k$ for some integer
k. Describe $f_*: \pi(S^1, 1) \to \pi(S^1, 1)$ in terms of the isomorphism
$\pi(S^1, 1) \cong \mathbb{Z}$.

(c) Let α, β be the following closed paths in $S^1 \times S^1$.

$$\alpha(t) = (\exp(2\pi it), 1), \qquad \beta(t) = (1, \exp(2\pi it).)$$

Show, by means of diagrams, that $\alpha * \beta \sim \beta * \alpha$.

(d) Calculate $\pi(\underbrace{S^1 \times S^1 \times ... \times S^1}_{n}, (1, 1, ..., 1))$.

(e) Using Exercise 15.16(c) deduce that the torus is not homeomorphic
to the sphere S^2.

(f) Prove that the set of points $z \in D^2$ for which $D^2 - \{ z \}$ is simply
connected is precisely S^2. Hence prove that if $f: D^2 \to D^2$ is a
homeomorphism then $f(S^1) = S^1$.

(g) Find the fundamental groups of the following spaces.
(i) $\mathbb{C}^* = \mathbb{C} - \{ 0 \}$;
(ii) $\mathbb{C}^*/G$, where G is the group of homeomorphisms $\{ \varphi^n; n \in \mathbb{Z} \}$ with $\varphi(z) = 2z$.
(iii) $\mathbb{C}^*/H$ where $H = \{ \psi^n; n \in \mathbb{Z} \}$ with $\psi(z) = 2\bar{z}$.
(iv) $\mathbb{C}^*/\{ e, a \}$, where e is the identity homeomorphism and
$az = -\bar{z}$.

Covering spaces

In this and the next few chapters we explore generalizations of the results and concepts of Chapter 16.

Let p: $\widetilde{X} \to X$ be a continuous map. We say that the open subset $U \subseteq X$ is *evenly covered* by p if $p^{-1}(U)$ is the disjoint union of open subsets of $\widetilde{X}$ each of which is mapped homeomorphically onto U by p. The continuous map p: $\widetilde{X} \to X$ is said to be a *covering map* if each point $x \in X$ has an open neighbourhood evenly covered by p. Then we say that p: $\widetilde{X} \to X$ is a *covering*, $\widetilde{X}$ is the *covering space* of X and X is the *base space* of the covering map p: $\widetilde{X} \to X$.

In other words p: $\widetilde{X} \to X$ is a covering if
  (i) p is onto, and
  (ii) for all $x \in X$ there is an open neighbourhood U of x such that

$$p^{-1}(U) = \bigcup_{j \in J} U_j$$

for some collection $\{ U_j; j \in J \}$ of subsets of $\widetilde{X}$ satisfying $U_j \cap U_k = \emptyset$ if $j \neq k$ and with $p|U_j: U_j \to U$ a homeomorphism for each $j \in J$.

From Chapter 16 we see that e: $\mathbb{R} \to S^1$ is a covering. Obviously any homeomorphism h: $X \to X$ is a covering map. Another trivial example of a covering p: $\widetilde{X} \to X$ is to let $\widetilde{X}$ be $X \times Y$ where Y is a discrete space and p is the canonical projection. An interesting example is $p_n: S^1 \to S^1$ where $p_n(z) = z^n$ ($n \neq 0$, think of $S^1 \subset \mathbb{C}$). To see that this is indeed a covering map just note that $S^1 - \{ x \}$ is evenly covered by p_n for all $x \in S^1$.

Certain G-spaces lead to covering spaces. Suppose that X is a G-space. We say that the action of G on X is *properly discontinuous* if for each $x \in X$ there is an open neighbourhood V of x such that $g \cdot V \cap g' \cdot V = \emptyset$ for all $g, g' \in G$ with $g \neq g'$. Notice that if the action is properly discontinuous then $g \cdot x \neq x$ for all $g \in G$, $g \neq 1$ and all $x \in X$, because if $x \in V$ then $g \cdot x \in g \cdot V$. Before giving examples we shall prove a theorem which explains the reason for introducing properly discontinuous actions.

17.1   Theorem

Let X be a G-space. If the action of G on X is properly discontinuous then p: $X \to X/G$ is a covering.

Proof First note that p: $X \to X/G$ is a continuous surjective map. Also, by Theorem 5.12, p is an open mapping. Let U be an open neighbourhood of $x \in X$ satisfying the condition of proper discontinuity. Since p is an open map, p(U) is an open neighbourhood of $G \cdot x = p(x)$ and $p^{-1}(p(U)) = \bigcup_{g \in G} g \cdot U$

(see the proof of Theorem 5.12) with $\{ g \cdot U; g \in G \}$ being disjoint open subsets of X. Furthermore $p|g \cdot U: g \cdot U \to p(U)$ is a continuous open bijective mapping and hence a homeomorphism.

The action of $\mathbb{Z}$ on $\mathbb{R}$ given by $x \to x + n$ is properly discontinuous since if $x \in \mathbb{R}$ and $\epsilon < \frac{1}{2}$ then $(x-\epsilon, x+\epsilon)$ is an open neighbourhood of x satisfying the required condition. Since this action of $\mathbb{Z}$ on $\mathbb{R}$ makes $\mathbb{R}$ into a $\mathbb{Z}$-space we see that p: $\mathbb{R} \to \mathbb{R}/\mathbb{Z}$ is a covering map. (The reader should check that this example is identical to e: $\mathbb{R} \to S^1$.)

The next example shows that the natural map $S^n \to \mathbb{R}P^n$ is a covering map. Consider the $\mathbb{Z}_2$-space S^n where $\mathbb{Z}_2$ acts as $\pm 1 \cdot x = \pm x$. For $x \in S^n$, the set

$$\{ y \in S^n; \|y - x\| < \frac{1}{2} \}$$

is an open neighbourhood of x which satisfies the condition for proper discontinuity. Alternatively, since $x \neq -x$ and S^n is Hausdorff, there exist disjoint open neighbourhoods V,W of x and $-x$ respectively. The neighbourhood $V \cap (-W)$ of x satisfies the condition of proper discontinuity. This example generalizes: Recall that a group G acts *freely* on X (or is a *free action*) if $g \cdot x \neq x$ for all $x \in X$ and $g \in G$, $g \neq 1$.

17.2   Theorem

If G is a finite group acting freely on a Hausdorff space X then the action of G on X is properly discontinuous.

Proof Let $G = \{ 1 = g_0, g_1, g_2, ..., g_n \}$. Since X is Hausdorff there are open neighbourhoods $U_0, U_1, ..., U_n$ of $g_0 \cdot x, g_1 \cdot x, ..., g_n \cdot x$ respectively with $U_0 \cap U_j = \emptyset$ for j = 1,2,...,n. Let U be the intersection $\bigcap_{j=0}^{n} g_j^{-1} \cdot U_j$, which is clearly an open neighbourhood of x.

Now $g_i \cdot U = \bigcap_{j=0}^{n} g_i \cdot (g_j^{-1} \cdot U_j) \subseteq U_i$ and

$$g_i \cdot U \cap g_j \cdot U = g_j \cdot ((g_j^{-1} g_i \cdot U) \cap U)$$
$$= g_j \cdot (g_k \cdot U \cap U) \qquad \text{(for some k)}$$
$$= \emptyset$$

since $g_k \cdot U \subseteq U_k$ and $U \subseteq U_0$. Hence the action of G on X is properly discontinuous.

A nice example of a free action is given by the cyclic group $\mathbb{Z}_p$ and the 3-sphere $S^3 \subset \mathbb{C}^2$,

$$S^3 = \{ (z_0, z_1) \in \mathbb{C}^2 ; |z_0|^2 + |z_1|^2 = 1 \} .$$

Let q be prime to p and define h: $S^3 \to S^3$ by

$$h(z_0, z_1) = (\exp(2\pi i/p)z_0, \exp(2\pi iq/p)z_1).$$

Then h is a homeomorphism of S^3 with $h^p = 1$. We let $\mathbb{Z}_p$ act on S^3 by

$$n \cdot (z_0, z_1) = h^n(z_0, z_1), n \in \mathbb{Z}_p = \{ 0, 1, ..., p-1 \} .$$

The action is free, S^3 is Hausdorff and so $S^3 \to S^3/\mathbb{Z}_p$ is a covering map. The quotient $S^3/\mathbb{Z}_p$ is called a *lens space* and is denoted by L(p,q). Note that L(2,1) is $\mathbb{R}P^3$. There is an obvious generalization of the above example to an action of $\mathbb{Z}_p$ on $S^{2n+1} \subset \mathbb{C}^{n+1}$; we leave the details for the reader.

It is time we had some general results about covering maps. Notice that in the examples arising from group actions the covering map is open and the base space has the quotient topology with respect to the covering map. This is in fact true for all coverings.

17.3 Theorem

Let p: $\widetilde{X} \to X$ be a covering map. Then
(i) p is an open map.
(ii) X has the quotient topology with respect to p.

Proof Let U be an open subset of $\widetilde{X}$ and let $x \in p(U)$. Since p is a covering map there is an evenly covered open neighbourhood V of x. Let $\widetilde{x} \in p^{-1}(x) \cap U$; since $\widetilde{x} \in p^{-1}(V) = \bigcup_{j \in J} V_j$ there is some open set V_j in $\widetilde{X}$ with $\widetilde{x} \in V_j$. Since $V_j \cap U$ is open in V_j and $p|V_j$ is a homeomorphism of V_j onto V we see that $p(V_j \cap U)$ is an open subset of V. But V being open in X means that $p(V_j \cap U)$ is open in X. Since $x \in p(V_j \cap U) \subseteq p(U)$ it follows that p(U) is open and hence p is an open map.

The second part of the theorem follows from the fact that p is a continuous open mapping and hence a subset V of X is open if and only if $p^{-1}(V)$ is open.

Many of the results that we proved in the last chapter for the covering e: $\mathbb{R} \to S^1$ generalize to other coverings. If p: $\tilde{X} \to X$ is a covering and f: $Y \to X$ is a continuous map then a *lift* of f: $Y \to X$ is a continuous map $\tilde{f}$: $Y \to \tilde{X}$ such that $p\tilde{f} = f$. The next result shows that if a lift exists then it is essentially unique.

17.4   Lemma

Let p: $\tilde{X} \to X$ be a covering and let $\tilde{f}, \tilde{\tilde{f}}$: $Y \to \tilde{X}$ be two liftings of f: $Y \to X$. If Y is connected and $\tilde{f}(y_0) = \tilde{\tilde{f}}(y_0)$ for some $y_0 \in Y$ then $\tilde{f} = \tilde{\tilde{f}}$.

Proof Define Y' to be

$$\{ y \in Y; \tilde{f}(y) = \tilde{\tilde{f}}(y) \}$$

which is non-empty since $y_0 \in Y'$ by assumption. We shall show that Y' is both open and closed. Let $y \in Y$; then there is an open neighbourhood V of f(y) which is evenly covered by p, i.e. $p^{-1}(V)$ is the disjoint union of $\{ V_j; j \in J \}$ and $p|V_j: V_j \to V$ is a homeomorphism for each $j \in J$. If $y \in Y'$ then $\tilde{f}(y) = \tilde{\tilde{f}}(y) \in V_k$ for some $k \in J$ and $\tilde{f}^{-1}(V_k) \cap \tilde{\tilde{f}}^{-1}(V_k)$ is an open neighbourhood of y contained in Y'. To see this let $x \in \tilde{f}^{-1}(V_k) \cap \tilde{\tilde{f}}^{-1}(V_k)$; then $\tilde{f}(x) \in V_k$ and $\tilde{\tilde{f}}(x) \in V_k$, but also $p\tilde{f}(x) = p\tilde{\tilde{f}}(x)$. Since $p|V_k$ is a homeomorphism it follows that $\tilde{f}(x) = \tilde{\tilde{f}}(x)$. Thus each point of Y' has an open neighbourhood contained in Y' and so Y' is open. On the other hand if $y \notin Y'$ then $\tilde{f}(y) \in V_k$ and $\tilde{\tilde{f}}(y) \in V_\ell$ for some k, ℓ with $k \neq \ell$. Hence $\tilde{f}^{-1}(V_k) \cap \tilde{\tilde{f}}^{-1}(V_\ell)$ is an open neighbourhood of y contained in the complement of Y' (argue as above). Thus Y' is closed. Since Y is connected it follows that Y = Y' and so $\tilde{f} = \tilde{\tilde{f}}$.

A curious corollary is:

17.5   Corollary

Suppose $\tilde{X}$ is path connected and φ: $\tilde{X} \to \tilde{X}$ is a continuous map with $p\varphi = p$. If $\varphi(x_1) = x_1$ for some $x_1 \in \tilde{X}$ then $\varphi(x) = x$ for all $x \in \tilde{X}$ (i.e. φ is the identity mapping).

Proof Let x be any point of $\tilde{X}$ and let α: $I \to \tilde{X}$ be a path from x_1 to x. Since $\varphi(x_1) = x_1$ the paths α and $\varphi\alpha$ both begin at x_1. Furthermore $p\alpha = p\varphi\alpha$ so that α and $\varphi\alpha$ are lifts of the path $p\alpha$: $I \to X$. By the above lemma $\alpha = \varphi\alpha$ and in particular the end points of α and $\varphi\alpha$ coincide, i.e. $\varphi(x) = x$.

The next result, known as the *homotopy path lifting theorem,* is proved in the same way as Theorem 16.4 and Lemma 16.5.

17.6   Theorem

Let p: $\tilde{X} \to X$ be a covering.

(i) Given a path f: $I \to X$ and a $\in \tilde{X}$ with p(a) = f(0) there is a unique path $\tilde{f}$: $I \to X$ such that $p\tilde{f} = f$ and $\tilde{f}(0) = a$.

(ii) Given a continuous map F: $I \times I \to X$ and a $\in \tilde{X}$ with p(a) = F(0,0), there is a unique continuous map $\tilde{F}$: $I \times I \to \tilde{X}$ such that $p\tilde{F} = F$ and $\tilde{F}(0,0) = a$.

As a corollary we have the *monodromy theorem,* the proof of which is identical to Corollary 16.6.

17.7   Corollary

Suppose that f_0 and f_1 are equivalent paths in X. If $\tilde{f}_0(0) = \tilde{f}_1(0)$ then $\tilde{f}_0(1) = \tilde{f}_1(1)$.

Continuing with generalizations of results in Chapter 16, we have the following result.

17.8   Theorem

Let p: $\tilde{X} \to X$ be a covering with $\tilde{X}$ being simply connected. Then there is a one-to-one correspondence between the sets $\pi(X,p(a))$ and $p^{-1}(p(a))$ where a $\in \tilde{X}$.

Proof The proof is essentially contained in the proof of Theorem 16.7. We shall give the salient points. First define

$$\varphi\colon \pi(X,p(a)) \to p^{-1}(p(a))$$

by

$$\varphi([f]) = \tilde{f}(1)$$

where $\tilde{f}$ is a lift of f with $\tilde{f}(0) = a$. This is well defined by Corollary 17.7.

Next, we shall define

$$\psi\colon p^{-1}(p(a)) \to \pi(X,p(a)).$$

To do this let $x \in p^{-1}(p(a))$ and choose a path f from a to x. Since $\tilde{X}$ is simply connected, any two such paths are equivalent, so that [pf] is a well-defined element of $\pi(X,p(a))$. Define $\psi(x) = [pf]$. It is easy to check that $\varphi\psi = 1$ and $\psi\varphi = 1$, so that φ and ψ are bijections.

The moral of the above result is that in order to calculate $\pi(X,x_0)$ we should find a covering p: $\widetilde{X} \to X$ so that the covering space $\widetilde{X}$ is simply connected. Then find a group structure on $p^{-1}(x_0)$ so that the bijection φ: $\pi(X,x_0) \to p^{-1}(x_0)$ is an isomorphism of groups. This is essentially what we did in Chapter 16. In general it is not easy to do; some special cases will appear in the subsequent chapters (see also Chapter 21).

17.9   Exercises

(a) Let p: $\widetilde{X} \to X$ be a covering map, let X_0 be a subset of X and let $\widetilde{X}_0 = p^{-1}(X_0)$. Prove that p_0: $\widetilde{X}_0 \to X_0$, given by $p_0(x) = p(x)$, is a covering map.

(b) Let $\widetilde{X} = \{ (x,y) \in \mathbb{R}^2 ; x \text{ or } y \text{ is an integer} \}$, Let $X = \{ (z_1,z_2) \in S^1 \times S^1 ; z_1 = 1 \text{ or } z_2 = 1 \}$ and let p: $\widetilde{X} \to X$ be defined by $p(x,y) = (\exp(2\pi ix), \exp(2\pi iy))$. Show that p: $\widetilde{X} \to X$ is a covering map.

(c) Which of the following are covering mappings?
   (i) p: $\mathbb{C}^* \to \mathbb{C}^*$ given by $p(z) = z^n$ where n is a fixed integer;
   (ii) sin: $\mathbb{C} \to \mathbb{C}$;
   (iii) p: $U \to \mathbb{C}^*$ given by $p(z) = (1-z)^m z^n$ where m,n are fixed integers and $U = \mathbb{C}^* - \{ 1 \}$.

(d) Let p: $\widetilde{X} \to X$ and q: $\widetilde{Y} \to Y$ be covering maps.
   (i) Prove that $p \times q$: $\widetilde{X} \times \widetilde{Y} \to X \times Y$ is a covering map.
   (ii) Prove that if X = Y and
   $$\widetilde{W} = \{ (\widetilde{x},\widetilde{y}) \in \widetilde{X} \times \widetilde{X}: p(\widetilde{x}) = q(\widetilde{y}) \}$$
   then f: $\widetilde{W} \to X$, defined by $f(\widetilde{x},\widetilde{y}) = p(\widetilde{x})$, is a covering map.
   (iii) Identify $\widetilde{W}$ and f when both p: $\widetilde{X} \to X$ and q: $\widetilde{Y} \to Y$ are e: $\mathbb{R} \to S^1$, where $e(t) = \exp(2\pi it)$.

(e) Let a: $\mathbb{C} \to \mathbb{C}$ and b: $\mathbb{C} \to \mathbb{C}$ be the homeomorphisms of the complex plane $\mathbb{C}$ defined by
   $$az = z + i,$$
   $$bz = \bar{z} + \tfrac{1}{2} + i.$$
   Show that $ba = a^{-1}b$ and deduce that
   $$G = \{ a^m b^{2n} b^\epsilon : m \in \mathbb{Z}, n \in \mathbb{Z}, \epsilon = 0 \text{ or } 1 \}$$
   is a group of homeomorphisms of $\mathbb{C}$. Furthermore, prove that the action of G is properly discontinuous and that the orbit space $\mathbb{C}/G$ is Hausdorff.

(f) (Continuation of (e).) Find a 'half-open rectangle' containing exactly one point from each orbit of G and hence show that $\mathbb{C}/G$ is a Klein bottle.

(g) (Continuation of (f).) *An embedding of the Klein bottle in* $\mathbf{R}^4$. Let $\varphi: \mathbf{C} \to \mathbf{R}^5$ be defined by

$$\varphi(x+iy) = (\cos(2\pi y), \cos(4\pi x), \sin(4\pi x), \sin(2\pi y) \cos(2\pi x),$$
$$\sin(2\pi x) \sin(2\pi y)).$$

Show that φ identifies to a point each of the orbits of the group G and deduce that $\mathbf{C}/G$ is homeomorphic to the image of φ.

Show that the restriction of $\psi: \mathbf{R}^5 \to \mathbf{R}^4$, where

$$\psi(p,q,r,s,t) = ((p+2)q, (p+2)r,s,t),$$

to the image of φ is a homeomorphism.

(h) Suppose p: $\widetilde{X} \to X$ is a covering map with X path connected. Prove that the cardinal number of $p^{-1}(x)$ is independent of $x \in X$. If this number is n then we say that p: $\widetilde{X} \to X$ is an n-*fold covering*.

(i) Find a two-fold covering p: $S^1 \times S^1 \to K$ where K is the Klein bottle.

(j) A subset Σ of a space is a *simple closed curve* if it is homeomorphic to S^1. Let p: $S^2 \to \mathbf{R}P^2$ be the canonical projection of the sphere onto the projective plane. Prove that if Σ is a simple closed curve in $\mathbf{R}P^2$ then $p^{-1}(\Sigma)$ is either a simple closed curve in S^2 or is a union of two disjoint simple closed curves. (Hint: Consider Σ as the image of a closed path in $\mathbf{R}P^2$.)

(k) Calculate $\pi(S^1 \times S^1, (1,1))$ directly from results in this chapter. (Hint: Using Exercises (d) and (i) above we have a covering map $\mathbf{R} \times \quad \to S^1 \times S^1$; then use Theorem 17.8.)

(l) Assuming that S^n is simply connected for $n \geq 2$ (Exercise 15.16(c)), show that the fundamental group of $\mathbf{R}P^n$ ($n \geq 2$) is cyclic of order 2. Furthermore show that if p is a prime number then the fundamental group of the lens space $L(p,q)$ is cyclic of order p.

(m) Does there exist a topological space Y such that $S^1 \times Y$ is homeomorphic to $\mathbf{R}P^2$ or to S^2?

(n) Suppose that p: $X \to Y$ is a covering map and that X,Y are both Hausdorff spaces. Prove that X is an n-manifold if and only if Y is an n-manifold.

(o) Let p: $\widetilde{X} \to X$ be a covering and let Y be a space. Suppose that f: $Y \to X$ has a lift $\widetilde{f}: Y \to \widetilde{X}$. Prove that any homotopy F: $Y \times I \to X$ with $F(y,0) = f(y)$, $y \in Y$, can be lifted to a homotopy $\widetilde{F}: Y \times I \to \widetilde{X}$ with $\widetilde{F}(y,0) = \widetilde{f}(y)$, $y \in Y$.

(p) Let p: $\widetilde{X} \to X$ be a covering and let f, g: $Y \to \widetilde{X}$ be two continuous maps with pf = pg. Prove that the set of points in Y for which f and g agree is an open and closed subset of Y.

(q) Let p: $\widetilde{X} \to X$ be a covering with X locally path connected (see

Exercise 12.10(j)). Prove that $\tilde{X}$ is also locally path connected.

(r) A *covering transformation* h of the covering p: $\tilde{X} \to X$ is a homeomorphism h: $\tilde{X} \to \tilde{X}$ for which ph = h. Prove that the set of covering transformations forms a group.

(s) Let p: $\tilde{X} \to X$ be a covering in which $\tilde{X}$ is connected and locally path connected. Prove that the action of the group of covering transformations of p: $\tilde{X} \to X$ on $\tilde{X}$ is properly discontinuous.

The fundamental group of a covering space

This chapter is concerned with $\pi(\widetilde{X}, \widetilde{x}_0)$ and its relation with $\pi(X, x_0)$, where $p: \widetilde{X} \to X$ is a covering and $p(\widetilde{x}_0) = x_0$. Most of the results are in the exercises.

The first result we give follows immediately from Theorem 17.6.

18.1 Theorem

If $p: \widetilde{X} \to X$ is a covering with $\widetilde{x}_0 \in \widetilde{X}$, $x_0 \in X$ such that $p(\widetilde{x}_0) = x_0$, then the induced homomorphism

$$p_*: \pi(\widetilde{X}, \widetilde{x}_0) \to \pi(X, x_0)$$

is a monomorphism.

A natural question to ask is what happens if we change the base points. This is answered next.

18.2 Theorem

Let $p: \widetilde{X} \to X$ be a covering with $\widetilde{X}$ path connected. If $\widetilde{x}_0, \widetilde{x}_1 \in \widetilde{X}$ then there is a path f in X from $p(\widetilde{x}_0)$ to $p(\widetilde{x}_1)$ such that

$$u_f p_* \, \pi(\widetilde{X}, \widetilde{x}_0) = p_* \, \pi(\widetilde{X}, \widetilde{x}_1).$$

Proof Let g be a path in $\widetilde{X}$ from $\widetilde{x}_0$ to $\widetilde{x}_1$. The path g determines an isomorphism u_g from $\pi(\widetilde{X}, \widetilde{x}_0)$ to $\pi(\widetilde{X}, \widetilde{x}_1)$, so that $u_g \, \pi(\widetilde{X}, \widetilde{x}_0) = \pi(\widetilde{X}, \widetilde{x}_1)$. Applying the homomorphism p_* gives

$$p_* u_g \, \pi(\widetilde{X}, \widetilde{x}_0) = p_* \, \pi(\widetilde{X}, \widetilde{x}_1).$$

But $p_* u_g = u_{pg} p_*$ (easy: see Exercise 15.11(c)), so that the path $f = pg$ satisfies the required conditions.

If, in the above theorem, $p(\widetilde{x}_0) = p(\widetilde{x}_1) = x_0$ then the path f determines an element $[f]$ of $\pi(X, x_0)$ and so

$$p_* \, \pi(\widetilde{X}, \widetilde{x}_1) = [f]^{-1} (p_* \, \pi(\widetilde{X}, \widetilde{x}_0)) [f].$$

In other words the subgroups $p_* \pi(\widetilde{X},\widetilde{x}_0)$ and $p_* \pi(\widetilde{X},\widetilde{x}_1)$ are conjugate subgroups of $\pi(X,x_0)$. In fact we can say more:

18.3   Theorem
Let $p: \widetilde{X} \to X$ be a covering with $\widetilde{X}$ path connected. For $x_0 \in X$ the collection

$$\{ p_* \pi(\widetilde{X},\widetilde{x}_0); \widetilde{x}_0 \in p^{-1}(x_0) \}$$

is a conjugacy class in $\pi(X,x_0)$.

Proof We have already shown that any two subgroups in the collection are conjugate. Suppose therefore that H is a subgroup of $\pi(X,x_0)$ which is conjugate to one of the subgroups $p_* \pi(\widetilde{X},\widetilde{x}_0)$. Thus

$$H = \alpha^{-1}(p_* \pi(\widetilde{X},\widetilde{x}_0))\alpha$$

for some $\alpha \in \pi(X,x_0)$. Let $\alpha = [f]$ and let $\widetilde{f}$ be a lift of f that begins at $\widetilde{x}_0$. We then have

$$p_* \pi(\widetilde{X},\widetilde{f}(1)) = u_f p_* \pi(\widetilde{X},x_0) = H$$

so that H belongs to the collection.

Other relations between $\pi(\widetilde{X},\widetilde{x}_0)$ and $\pi(X,x_0)$ will be given as exercises.

18.4   Exercises
Throughout these exercises let $p: \widetilde{X} \to X$ be a covering with $\widetilde{X}$ path connected and $x_0 \in X$.

(a) For $\widetilde{x} \in p^{-1}(x_0)$ and $[f] \in \pi(X,x_0)$ define $\widetilde{x} \cdot [f]$ by

$$\widetilde{x} \cdot [f] = \widetilde{f}(1)$$

where $\widetilde{f}$ is the unique lift of f that begins at $\widetilde{x}$. Prove that this defines a right action of the group $\pi(X,x_0)$ on the set $p^{-1}(x_0)$. (Hint: Look at the proof of Theorem 16.7 and use the notation $\ell_a(f)$ for the lift of f that begins at a.)

(b) We say that a group G acts *transitively* on a set S if for all $a,b \in S$ there is an element $g \in G$ such that $g \cdot a = b$; in other words $S = G \cdot a$, the orbit of a, for $a \in S$. Prove that $\pi(X,x_0)$ acts transitively on $p^{-1}(x_0)$.

(c) Prove that there is a $\pi(X,x_0)$ equivariant bijection between $p^{-1}(x_0)$ and the set of right cosets of $p_* \pi(\widetilde{X},\widetilde{x}_0)$ in $\pi(X,x_0)$. (Hint: Use Exercise 5.9(d) with the word 'right' substituted for 'left' and show that the stabilizer of the action of $\pi(X,x_0)$ on $p^{-1}(x_0)$ is $p_* \pi(\widetilde{X},\widetilde{x}_0)$.)

(d) Deduce from (c) above that if $\widetilde{X}$ is simply connected then there is a $\pi(X,x_0)$ equivariant bijection between $p^{-1}(x_0)$ and $\pi(X,x_0)$.

(e) Show that if p: $\widetilde{X} \to X$ is an n-fold covering (i.e. $p^{-1}(x_0)$ consists of n points) then

$$p_* \ \pi(\widetilde{X},\widetilde{x}_0) \to \pi(X,x_0)$$

is the inclusion of a subgroup of index n.

(f) Suppose that the fundamental group of X is $\mathbb{Z}$ and $p^{-1}(x_0)$ is finite. Find the fundamental group of $\widetilde{X}$.

(g) Prove that if X is simply connected then p is a homeomorphism.

(h) Suppose that $\widetilde{X} = X$. Prove that p is a homeomorphism if the fundamental group of X is finite. Is p necessarily a homeomorphism if the fundamental group of X is not finite?

(i) A covering is said to be *regular* if for some $\widetilde{x}_0 \in \widetilde{X}$ the group $p_* \pi(\widetilde{X},\widetilde{x}_0)$ is a normal subgroup of $\pi(X,x_0)$. Prove that if f is a closed path in X then either every lifting of f is closed or none is closed.

(j) Suppose that p: $\widetilde{X} \to X$ is a covering obtained from a properly discontinuous action of G on $\widetilde{X}$ (i.e. $X = \widetilde{X}/G$). Prove that p: $\widetilde{X} \to X$ is regular.

(k) Prove that p is a homeomorphism if and only if

$$p_* \ \pi(\widetilde{X},\widetilde{x}_0) = \pi(X,x_0).$$

19

The fundamental group of an orbit space

Throughout this chapter we shall assume that X is a path connected space with a properly discontinuous action of G on it. Thus $p: X \to X/G$ is a covering. The object of this chapter is to find the relation between G and the fundamental group of the orbit space X/G.

Let $x_0 \in X$ and $y_0 = p(x_0) \in X/G$. Notice that

$$p^{-1}(y_0) = \{\, g{\cdot}x_0 ; g \in G \,\}.$$

If $[f] \in \pi\,(X/G, y_0)$ then there is a unique lift $\tilde{f}$ of f that begins at $x_0 \in X$. The element $\tilde{f}(1) \in p^{-1}(y_0)$ and so there is a unique element $g_f \in G$ such that $\tilde{f}(1) = g_f \cdot x_0$. The correspondence $f \to g_f$ therefore defines a function

$$\varphi: \pi(X/G, y_0) \to G.$$

19.1   Theorem
The function $\varphi: \pi(X/G, y_0) \to G$ is a homomorphism of groups.

Proof Consider two closed paths f, f' in X/G based at y_0. If $\widetilde{f * f'}$ is the unique lift of $f * f'$ that begins at $x_0 \in X$ then

$$\widetilde{f * f'} = \tilde{f} * \ell_a (f')$$

where $\tilde{f}$ is the unique lift of f that begins at x_0, and $\ell_a(f')$ is the unique lift of f' that begins at $a = \tilde{f}(1)$. This is because $\tilde{f} * \ell_a(f')$ is also a lift of $f * f'$ that begins at $x_0 \in X$. Let $\tilde{f}'$ be the unique lift of f' that begins at x_0. Since $g_f \cdot \tilde{f}'$ is a lift of f' that begins at $g_f \cdot x_0$ and $a = \tilde{f}(1) = g_f \cdot x_0$ it follows that $\ell_a(f') = g_f \cdot \tilde{f}'$. Hence

$$\widetilde{f * f'}\,(1) = g_f \cdot \tilde{f}'\,(1)$$
$$= g_f \cdot (g_{f'} \cdot x_0)$$
$$= (g_f g_{f'}) \cdot x_0.$$

It follows that $\varphi\,([f]\,[f']) = (\varphi\,[f])(\varphi\,[f'])$ and so φ is a homomorphism.

The kernel of the homomorphism φ is given next.

19.2   Lemma
The kernel of $\varphi: \pi(X/G, y_0) \to G$ is the subgroup $p_* \pi(X, x_0)$.

Proof The kernel of φ is the set of elements $[f] \in \pi(X/G, y_0)$ such that $\varphi[f] = 1$. This is precisely those elements $[f] \in \pi(X/G, y_0)$ for which $\tilde{f}(1) = x_0$, i.e. for which $\tilde{f}$ is a closed path in X based at $x_0 \in X$. Thus it is the set of elements $[f] \in \pi(X/G, y_0)$ of the form $[p\tilde{f}]$ for $[\tilde{f}] \in \pi(X, x_0)$, i.e. $p_* \pi(X, x_0)$.

In particular, $p_* \pi(X, x_0)$ is a normal subgroup of $\pi(X/G, y_0)$ and so the quotient group

$$\pi(X/G, y_0)/p_* \pi(X, x_0)$$

is defined.

19.3   Theorem
The groups $\pi(X/G, y_0)/p_* \pi(X, x_0)$ and G are isomorphic.

Proof We just need to show that the homomorphism

$$\varphi: \pi(X/G, y_0) \to G$$

is surjective. If $g \in G$ let f_g be a path in X from x_0 to $g \cdot x_0$. This determines an element $[pf_g] \in \pi(X/G, y_0)$. By definition $\varphi[pf_g] \cdot x_0 = \widetilde{pf_g}(1)$ where $\widetilde{pf_g}$ is the unique lift of pf_g that begins at x_0. But f_g is such a lift and $f_g(1) = g \cdot x_0$; thus $\varphi[pf_g] = g$, showing that φ is surjective.

19.4   Corollary
If X is simply connected then $\pi(X/G, y_0) \cong G$.

From this corollary we can recapture the result of Chapter 16, namely that the fundamental group of the circle is $\mathbb{Z}$. This follows because S^1 is homeomorphic to $\mathbb{R}/\mathbb{Z}$. In the same way we can deduce that the fundamental group of $(S^1)^n \cong \mathbb{R}^n/\mathbb{Z}^n$ is isomorphic to $\mathbb{Z}^n$. Other examples, often based on previous exercises, are given in the next set of exercises.

19.5   Exercises
   (a) Show that the fundamental group of the lens space $L(p,q)$ is isomorphic to $\mathbb{Z}_p$. (Assume by Exercise 15.16(c) that S^3 is simply connected.)
   (b) Show that for any finitely generated abelian group G there exists a space X_G whose fundamental group is G. (Assume the fact that G

is a product of a number of copies of the integers $\mathbb{Z}$ and a number of cyclic groups.)

(c) Let $Y = \mathbb{C}^*/K$ where $\mathbb{C}^* = \mathbb{C} - \{0\}$ and K is the group of homeomorphisms $\{\varphi^n; n \in \mathbb{Z}\}$ with $\varphi(z) = 4z$. Prove, using Corollary 19.4, that the fundamental group of Y is $\mathbb{Z} \times \mathbb{Z}$. (Hint: Find a space X, a group G and a normal subgroup H so that X is a G-space with X simply connected, $X/H \cong \mathbb{C}^*$ and $G/H = K$. Then use Exercise 5.13(c).)

(d) Prove that $Tz = \bar{z} + 1 + i$ defines a homeomorphism $T: X \rightarrow X$ where $X = \mathbb{R} \times [0,1] \subset \mathbb{C}$. Show that if G is the group of homeomorphisms generated by T then X/G is the Möbius strip. Deduce that the fundamental group of the Möbius strip is $\mathbb{Z}$.

(e) Prove that the fundamental group of a Klein bottle is

$$G = \{ a^m b^{2n+\epsilon}; m \in \mathbb{Z}, n \in \mathbb{Z}, \epsilon = 0 \text{ or } 1, ba = a^{-1}b \},$$

i.e. G is the group on two generators a, b with one relation $ba = a^{-1}b$.

(f) Suppose that X has a properly discontinuous G-action. Recall from Exercise 18.4(a) that $\pi(X/G, y_0)$ acts on $p^{-1}(y_0)$ (on the right). Prove that

$$(g \cdot x) \cdot [f] = g \cdot (x \cdot [f])$$

for $g \in G$, $x \in p^{-1}(y_0)$ and $[f] \in \pi(X/G, y_0)$.

(g) Suppose that G, H are groups acting on a set S with G acting on the left and H acting on the right. Suppose also that

$$(g \cdot x) \cdot h = g \cdot (x \cdot h)$$

for all $g \in G$, $x \in S$, $h \in H$. Prove that if G acts freely and transitively on S then there is a homomorphism

$$\varphi: H \rightarrow G.$$

Furthermore show that the kernel of φ is the stabilizer under H of the point $x_0 \in S$ where $S = \{g \cdot x_0; g \in G\}$. (Hint: Define $\varphi(h)$, $h \in H$, to be the unique element of g such that $g \cdot x_0 = x_0 \cdot h$.)

(h) Use (f) and (g) above to reprove Theorem 19.1 and Lemma 19.2.

(i) Use (h) above and Exercise 18.4(c) to reprove Theorem 19.3.

The Borsuk–Ulam and ham-sandwich theorems

We shall give a few applications of the results from the preceding chapters. They are generalizations of the results in Chapter 10 and are all based upon the *Borsuk–Ulam theorem* (proved by K. Borsuk in the early 1930s after a conjecture by S. Ulam).

20.1 Theorem

There does not exist any continuous map $\varphi\colon S^2 \to S^1$ such that $\varphi(-x) = -\varphi(x)$.

This generalizes the result that there does not exist any continuous map $\varphi\colon S^1 \to S^0$ such that $\varphi(-x) = -\varphi(x)$ (such a map would be surjective, but S^1 is connected and S^0 is not). In fact there is a more general result: there does not exist any continuous map $\varphi\colon S^n \to S^{n-1}$, for $n \geq 1$, such that $\varphi(-x) = -\varphi(x)$. For $n > 2$ the proof is beyond the scope of this book; it relies for example on the higher homotopy groups.

To prove Theorem 20.1 we suppose that there does exist a continuous map $\varphi\colon S^2 \to S^1$ such that $\varphi(-x) = -\varphi(x)$. The group $\mathbb{Z}_2 = \{\pm 1\}$ acts antipodally on S^2 and on S^1 (i.e. $\pm 1 \cdot x = \pm x$); in each case the action is properly discontinuous. If $p_2\colon S^2 \to S^2/\mathbb{Z}_2$ and $p_1\colon S^1 \to S^1/\mathbb{Z}_2$ denote the canonical projections then φ induces a continuous map $\psi\colon S^2/\mathbb{Z}_2 \to S^1/\mathbb{Z}_2$ such that $p_1\varphi = \psi p_2$, namely $\psi(\{\pm x\}) = \{\pm\varphi(x)\}$; see the proof of Theorem 5.5. Let $a = (1,0,0) \in S^2$, where as usual

$$S^2 = \{(x, y, z) \in \mathbb{R}^3 ; x^2 + y^2 + z^2 = 1\}.$$

Also, let $b = p_2(a) \in S^2/\mathbb{Z}_2$. If f is the path in S^2 from a to $-a$ given by

$$f(t) = (\cos(\pi t), \sin(\pi t), 0) \qquad 0 \leq t \leq 1,$$

then $p_2 f$ is a closed path in $S^2/\mathbb{Z}_2$ based at b. We claim that the element $[p_2 f] \in \pi(S^2/\mathbb{Z}_2, b)$ satisfies

$$[p_2 f]^2 = [\epsilon_b].$$

$$(p_2 f * p_2 f)(t) = \begin{cases} p_2(\cos(2\pi t), \sin(2\pi t), 0) & 0 \leq t \leq \tfrac{1}{2}, \\ p_2(\cos(\pi(2t-1)), \sin(\pi(2t-1)), 0) & \tfrac{1}{2} \leq t \leq 1, \end{cases}$$

$$= \quad p_2(\cos(2\pi t), \sin(2\pi t), 0) \qquad 0 \leq t \leq 1.$$

Then define $F: I \times I \to S^2$ by

$$F(t,s) = (s + (1-s)\cos(2\pi t), (1-s)\sin(2\pi t), \sqrt{(2s(1-s)}$$
$$(1 - \cos(2\pi t))).$$

We see that $p_2 F: I \times I \to S^2/\mathbb{Z}_2$ is a continuous map satisfying

$$p_2 F(t,0) = (p_2 f * p_2 f)(t),$$
$$p_2 F(t,1) = p_2(1,0,0) = \epsilon_b(t),$$
$$p_2 F(0,s) = p_2(1,0,0) = p_2 F(1,s),$$

which shows that

$$[p_2 f]^2 = [\epsilon_b] \in \pi(S^2/\mathbb{Z}_2, b).$$

The map $\psi: S^2/\mathbb{Z}_2 \to S^1/\mathbb{Z}_2$ induces a homomorphism

$$\psi_*: \pi(S^2/\mathbb{Z}_2, b) \to \pi(S^1/\mathbb{Z}_2, \psi(b))$$

so that

$$[\psi p_2 f]^2 = [\epsilon_{\psi(b)}] \in \pi(S^1/\mathbb{Z}_2, \psi(b)).$$

We have of course $S^1/\mathbb{Z}_2 \cong S^1$ and

$$\pi(S^1/\mathbb{Z}_2, \psi(b)) = \{ \alpha^n; n \in \mathbb{Z} \}$$

for some $\alpha \in \pi(S^1/\mathbb{Z}_2, \psi(b))$. (In fact $\alpha = [p_1 g]$ where g: $I \to S^1$ is given by $g(t) = \exp(\pi i t) \in S^1 \subset \mathbb{C}$.) The fact that $[\psi p_2 f]^2 = [\epsilon_{\psi(b)}]$ means therefore that $[\psi p_2 f] = [\epsilon_{\psi(b)}]$, ($[\psi p_2 f] = \alpha^k$ for some k and $\alpha^{2k} = \alpha^0$ implies that $\alpha^k = \alpha^0$).

Remark: Exercise 15.16(c) tells us that S^2 is simply connected (see also Corollary 23.9), so that from the previous chapter we see that the fundamental group of $S^2/\mathbb{Z}_2$ is $\mathbb{Z}_2$. Thus ψ_* is a homomorphism from $\mathbb{Z}_2$ to $\mathbb{Z}$, but all such homomorphisms are trivial so that $[\psi p_2 f] = [\epsilon_{\psi(b)}] \in \pi(S^1/\mathbb{Z}_2, \psi(b))$. The previous argument was given to avoid using results from exercises or results to be proved later on.

To continue with the proof of Theorem 20.1, consider the results of Chapter 17 and look at the unique lifts of $\psi p_2 f$ and $\epsilon_{\psi(b)}$ to S^1 beginning at $\varphi(a)$. These are φf and $\epsilon_{\varphi(a)}$ respectively (remember $\psi p_2 = p_1 \varphi$). But $\varphi f(1) = \varphi(-a) = -\varphi(a)$ while $\epsilon_{\varphi(a)}(1) = \varphi(a)$ which contradicts the fact that $[\psi p_2 f] = [\epsilon_{\psi(b)}]$ and hence shows that φ does not exist. (Note also that $[p_2 f] \neq [\epsilon_b] \in \pi(S^2/\mathbb{Z}_2, b)$.)

20.2   Corollary

Let f: $S^2 \to \mathbb{R}^2$ be a continuous map such that $f(-x) = -f(x)$ for all $x \in S^2$. Then there exists a point $x \in S^2$ such that $f(x) = 0$.

Proof Suppose $f(x) \neq 0$ for all $x \in S^2$ and define g: $S^2 \to S^1$ by $g(x) = f(x)/\|f(x)\|$. The map g is continuous and $g(-x) = -g(x)$ which contradicts Theorem 20.1.

20.3   Corollary

Let f: $S^2 \to \mathbb{R}^2$ be a continuous map. Then there is a point $x \in S^2$ such that $f(x) = f(-x)$.

Proof If $f(x) \neq f(-x)$ for all $x \in S^n$ then we may define g: $S^2 \to \mathbb{R}^2$ by $g(x) = f(x) - f(-x)$, which is continuous and satisfies $g(-x) = -g(x)$ and $g(x) \neq 0$ for all $x \in S^n$. This contradicts Corollary 20.2.

Corollary 20.3 generalizes Corollary 10.3. Both of the above results do in fact also hold if S^2, $\mathbb{R}^2$ are replaced by S^n, $\mathbb{R}^n$.

Corollary 20.3 tells us in particular that there is no continuous injective map from S^2 to $\mathbb{R}^2$. This immediately gives the next result.

20.4   Corollary

No subset of $\mathbb{R}^2$ is homeomorphic to S^2.

As in Chapter 10, we have a physical interpretation.

20.5   Corollary

At any given moment of time there exists a pair of antipodal points on the surface of the earth which simultaneously have the same temperature and pressure.

The analogue of the first pancake problem is the *ham-sandwich theorem* which states that it is possible to cut a three layer sandwich consisting of bread, butter and ham (the ingredients are not really relevant) exactly in half with one stroke of a knife. More precisely:

20.6   Theorem

Let A, B, and C be bounded subsets of $\mathbb{R}^3$. Then there is a plane in $\mathbb{R}^3$ which divides each region exactly in half by volume.

Proof The proof is quite similar to the proof of Theorem 10.5. We may suppose that A, B, C lie within S, the sphere in $\mathbb{R}^3$ of diameter 1 and centre

0. For $x \in S$ let D_x denote the diameter line of S through x. For $t \in I$ let P_t denote the plane that is perpendicular to D_x and passes through the point on D_x at a distance t from x. Now P_t divides A into two parts A_1 and A_2 with A_1 closer to x than A_2. Define functions f_1, f_2 by

$$f_1(t) = \text{volume}(A_1), \qquad f_2(t) = \text{volume}(A_2).$$

Obviously f_1 and f_2 are continuous functions from I to $\mathbb{R}$ with f_1 monotone increasing and f_2 monotone decreasing. Therefore the function f: $I \to \mathbb{R}$ defined by $f(t) = f_1(t) - f_2(t)$ is continuous and monotone increasing. Furthermore $f(0) = -f(1)$ so that by the intermediate value theorem there is some $t \in I$ such that $f(t) = 0$. As f is monotone increasing it either vanishes at a single point a or on a closed interval [a,b]. In the former case we denote the single point a by $\alpha(x)$ while in the latter case we denote $(a + b)/2$ by $\alpha(x)$. Thus $P_{\alpha(x)}$ divides A into two equal parts. Note that α: $S \to \mathbb{R}$ is a continuous map that satisfies $\alpha(x) = 1 - \alpha(-x)$.

In a similar fashion we can define continuous functions β, γ: $S \to \mathbb{R}$ with $\beta(x) = 1 - \beta(-x)$, $\gamma(x) = 1 - \gamma(-x)$ and with the property that $P_{\beta(x)}, P_{\gamma(x)}$ divide B, C respectively exactly in half. Using the functions α, β, γ we now define φ: $S \to \mathbb{R}^2$ by

$$\varphi(x) = (\alpha(x) - \beta(x), \alpha(x) - \gamma(x))$$

Since α, β and γ are continuous so is φ. Furthermore $\varphi(-x) = -\varphi(x)$ so that by Corollary 20.2 there exists some point $y \in S$ such that $\varphi(y) = 0$. But this means that $\alpha(y) = \beta(y) = \gamma(y)$, so that the plane $P_{\alpha(y)}$ divides each of A, B and C exactly in half by volume.

20.7   Exercises

(a) Prove that if $n \geq 2$ then there does not exist any continuous map φ: $S^n \to S^1$ such that $\varphi(-x) = -\varphi(x)$.

(b) Let the group $\mathbb{Z}_p$ act on $S^3 \subset \mathbb{C}^2$ and $S^1 \subset \mathbb{C}$ as follows:

$$k \cdot (z_1, z_2) = (\exp(2\pi ik/p)z_1, \exp(2\pi ikq/p)z_2),$$
$$k \cdot z = \exp(2\pi ik/p)z,$$

where $k \in \mathbb{Z}_p = \{0, 1, ..., p-1\}$ and q is an integer prime to p. Prove that there does not exist any $\mathbb{Z}_p$ equivariant continuous map from S^3 to S^1.

(c) Is there an analogue in $\mathbb{R}^3$ of the second pancake problem?

(d) Suppose that X and Y are G-spaces for which the action of G is properly discontinuous. Suppose also that φ: $X \to Y$ is a G equivariant continuous map. Let ψ: $X/G \to Y/G$ denote the map induced by φ. Prove that the homomorphism ψ_*: $\pi(X/G, p(x_0))$

$\rightarrow \pi(Y/G, q\varphi(x_0))$ induces a homomorphism

$$\pi(X/G, p(x_0))/p_* \, \pi(X, x_0) \rightarrow \pi(Y/G, q\varphi(x_0))/q_* \, \pi(Y, \varphi(x_0))$$

which is an *isomorphism*, where $p: X \rightarrow X/G$ and $q: Y \rightarrow Y/G$ are the canonical projections.

(e) Use (d) above to reprove the Borsuk–Ulam theorem and to reprove the result in (b) above (a one line proof for each).

21

More on covering spaces: lifting theorems

Let p: $\widetilde{X} \to X$ be a covering and let f: $Y \to X$ be a continuous map with Y connected. Recall that in Chapter 17 we showed that if a lift $\widetilde{f}$ of f: $Y \to X$ exists then it is unique (essentially). Now, if the lift $\widetilde{f}$ exists then we have a commutative diagram

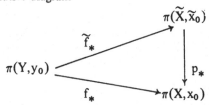

The homomorphism p_* is a monomorphism (Theorem 18.1) and $p_*\widetilde{f}_* = f_*$, so that

$$f_* \, \pi(Y,y_0) = p_*\widetilde{f}_* \, \pi(Y,y_0) \subseteq p_* \, \pi(\widetilde{X},\widetilde{x}_0).$$

Thus a necessary algebraic condition for the lift $\widetilde{f}$ to exist is that $f_*\pi(Y,y_0)$ $\subseteq p_* \, \pi(\widetilde{X},\widetilde{x}_0)$. It turns out that this is also a sufficient condition so long as we put an extra condition on Y. Thus a purely topological question is equivalent to a purely algebraic question. The condition we need on Y is that it is connected and locally path connected. A space Y is said to be *locally path connected* if for all y $\in$ Y every open neighbourhood of y contains a path connected open neighbourhood of y; see Exercise 12.10(j). A space that is connected and locally path connected is also path connected; we prove this now.

21.1 Lemma
If Y is connected and locally path connected then Y is path connected.

Proof Let y be some point of Y and let U be the set of points in Y which can be joined to y by a path in Y. If u $\in$ U then u has a path connected open neighbourhood V (because Y is an open neighbourhood of u and Y is locally path connected). If $v \in V$ then there is a path in V from u to v and a path in Y from y to u, so that there is a path from y to v in Y. Hence $V \subseteq U$. This shows that U is open. In a similar way we can show that Y - U is open so that U is open and closed. Since $y \in U$, the subset is non-empty. But Y is connected and so U must be Y, which proves that Y is path connected.

21.2   Theorem

Let p: $\widetilde{X} \to X$ be a covering, let Y be a connected and locally path connected space and let $y_0 \in Y$, $\widetilde{x}_0 \in \widetilde{X}$, $p(\widetilde{x}_0) = x_0$. Given a continuous map f: $Y \to X$ with $f(y_0) = x_0$, there exists a lift $\widetilde{f}$: $Y \to \widetilde{X}$ with $\widetilde{f}(y_0) = \widetilde{x}_0$ if and only if

$$f_* \, \pi(Y, y_0) \subseteq p_* \, \pi(\widetilde{X}, \widetilde{x}_0).$$

Proof We have already seen that the condition is necessary, thus it remains to show that it is sufficient. Suppose therefore that $f_* \, \pi(Y, y_0) \subseteq p_* \, \pi(\widetilde{X}, \widetilde{x}_0)$; we shall show that a lift $\widetilde{f}$ exists. The definition of $\widetilde{f}$ is as follows. Let $y \in Y$ and let φ: $I \to Y$ be a path in Y from y_0 to y. Thus $f\varphi$ is a path in X from x_0 to $f(y)$. By the homotopy path lifting theorem (Theorem 17.6(a)) there is a unique path $\widetilde{f\varphi}$: $I \to \widetilde{X}$ such that $\widetilde{f\varphi}(0) = x_0$ and $p \, \widetilde{f\varphi} = f\varphi$. We define $\widetilde{f}(y)$ to be $\widetilde{f\varphi}(1)$; see Figure 21.1.

Surprising though it may be, under this definition, $\widetilde{f}$ is well defined and it is continuous. First we show that $\widetilde{f}$ is well defined. The only choice we made was the path φ from y_0 to y, so let ψ be another path in Y from y_0 to y. The product path $\varphi * \bar{\psi}$ is a closed path in Y based at y_0. We have

$$f_* \, [\varphi * \bar{\psi}] = [f\varphi * f\bar{\psi}] \in f_* \, \pi(Y, y_0).$$

Figure 21.1

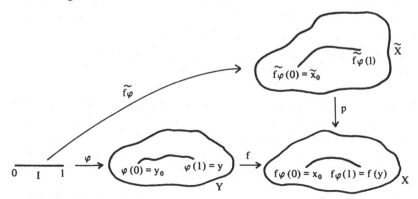

But $f_* \pi(Y,y_0) \subseteq p_* \pi(\widetilde{X},\widetilde{x}_0)$ and so there is a closed path α in $\widetilde{X}$ based at $\widetilde{x}_0$ such that

$$[f\varphi * f\overline{\psi}] = [p\alpha].$$

Using the results of Chapter 14 we have

$$f\varphi \sim f\varphi * \epsilon_{x_0}$$
$$\sim f\varphi * (f\overline{\psi} * f\psi)$$
$$\sim (f\varphi * f\overline{\psi}) * f\psi$$
$$\sim p\alpha * f\psi.$$

The path α is closed, thus

$$\widetilde{p\alpha * f\psi} = \alpha * \widetilde{f\psi}$$

and by the monodromy theorem (Theorem 17.7)

$$\widetilde{f\varphi}(1) = \widetilde{p\alpha * f\psi}(1) = (\alpha * \widetilde{f\psi})(1) = \widetilde{f\psi}(1)$$

which proves that $\widetilde{f}(y)$ is well defined. Notice that to define $\widetilde{f}$ we only need that Y is path connected.

To prove that $\widetilde{f}$ is continuous we need the extra assumption that Y is locally path connected. Suppose that U is an open subset of $\widetilde{X}$. Let $y \in \widetilde{f}^{-1}(U)$, thus U is an open neighbourhood of $\widetilde{f}(y)$. Let U′ be an evenly covered neighbourhood of $p\widetilde{f}(y) = f(y)$ such that $U' \subseteq p(U)$. By definition $p^{-1}(U') = \underset{j\in J}{\cup} V_j$ with each V_j being homeomorphic to U′; also $\widetilde{f}(y) \in V_k$ for some k. Since V_k and U are open neighbourhoods of $\widetilde{f}(y)$ so is $W = V_k \cap U$. Note that p(W) is evenly covered since U′ is and $p(W) \subseteq U'$. The map f is continuous and so $f^{-1}(p(W))$ is an open subset of Y which is an open neighbourhood of y. Because Y is locally path connected there is a path connected open neighbourhood V of y with $V \subseteq f^{-1}(p(W))$. We claim that $\widetilde{f}(V) \subseteq U$. Certainly $\widetilde{f}(y) \in U$. If y′ is another point of V then there is a path φ in V from y to y′ and by the definition of $\widetilde{f}$ we see that $\widetilde{f}(y')$ is $\widetilde{f\varphi}(1)$ where $\widetilde{f\varphi}$ is the unique lift of $f\varphi$ that begins at $\widetilde{f}(y)$. (If ψ is a path from y_0 to y then $\psi * \varphi$ is a path from y_0 to y′ and $\widetilde{f(\psi * \varphi)}(1) = \widetilde{f\varphi}(1)$ by the homotopy path lifting theorem.) The path $f\varphi$ has its image in $f(V) \subseteq p(W)$ and so the path $\widetilde{f\varphi}$ has its image in $p^{-1}(p(W))$. But $p^{-1}(p(W)) = \underset{j\in J}{\cup} W_j$ with the W_j being pairwise disjoint, each W_j being homeomorphic to p(W) and with one of the W_j, say W_k, being W. Since $\widetilde{f\varphi}(0) = \widetilde{f}(y) \in W$ it follows that $\widetilde{f\varphi}(1) = \widetilde{f}(y') \in W$ also. This proves that $\widetilde{f}(V) \subseteq W \subseteq U$ and hence $V \subseteq \widetilde{f}^{-1}(U)$. Thus every point of $\widetilde{f}^{-1}(U)$ has an open neighbourhood in $\widetilde{f}^{-1}(U)$, so that $\widetilde{f}^{-1}(U)$ is open and hence $\widetilde{f}$ is continuous.

We remark again that for $\widetilde{f}$ to be defined we just require that Y is path connected. For the continuity of $\tilde{f}$ we need that Y is locally path connected. We give an example now to show that if Y is path connected but not locally path connected then $\tilde{f}$ is not continuous. Let Y be the following subset of $\mathbb{R}^2$:

$$Y = A \cup B \cup C$$

where

$$A = \{ (x,y); x^2 + y^2 = 1, y \geq 0 \},$$
$$B = \{ (x,y); -1 \leq x \leq 0, y = 0 \},$$
$$C = \{ (x,y); 0 < x \leq 1, y = \tfrac{1}{2} \sin(\pi/x) \}.$$

See Figure 21.2; this space is called the *Polish circle*.

Figure 21.2

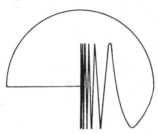

It is clear that Y is path connected. Furthermore since $B \cup C$ is not path connected it follows that Y is simply connected and that Y is not locally path connected. Consider the covering e: $\mathbb{R} \to S^1$ and let f: $Y \to S^1$ be the map defined by

$$f(x,y) = \begin{cases} (x,y) & \text{if } (x,y) \in A \subseteq Y, \\ (x, -\sqrt{(1-x^2)}) & \text{if } (x,y) \in B \cup C \subseteq Y. \end{cases}$$

Clearly f is continuous. Letting $y_0 = (1,0)$, $\tilde{x}_0 = 0$ we know that the condition

$$f_* \pi(Y,y_0) \subseteq p_* \pi(\tilde{X}, \tilde{x}_0)$$

is satisfied (where p=e, $\tilde{X}$= $\mathbb{R}$). We can define $\tilde{f}$ as in the proof of Theorem 21.2 to obtain

$$\tilde{f}(x,y) = \begin{cases} (\text{arc } \cos(x))/2\pi & \text{if } (x,y) \in A \cup B, \\ (\text{arc } \cos(x))/2\pi - 1 & \text{if } (x,y) \in C. \end{cases}$$

Clearly $p\tilde{f}=f$, $\tilde{f}(1,0) = 0$ but $\tilde{f}$ is not continuous at $(0,0) \in Y$.

There are several corollaries to Theorem 21.2. The first requires no proof.

21.3   **Corollary**

     If Y is simply connected and locally path connected then any continuous map f: Y → X lifts to $\tilde{f}$: Y → $\tilde{X}$.

21.4   **Corollary**

     Let p_1: X_1 → X, p_2: X_2 → X be two coverings with X_1, X_2 both connected and locally path connected. Let x_1, x_2, x_0 be base points of X_1, X_2, X respectively with $p_1(x_1) = p_2(x_2) = x_0$. If

$$p_{1_*} \pi(X_1, x_1) = p_{2_*} \pi(X_2, x_2)$$

then there is a base point preserving homeomorphism h: X_1 → X_2 such that $p_2 h = p_1$.

Proof Both p_1 and p_2 lift to maps $\tilde{p}_1$ and $\tilde{p}_2$ such that $p_2\tilde{p}_1 = p_1$ and $p_1\tilde{p}_2 = p_2$. Let $\varphi = \tilde{p}_2\tilde{p}_1$: X_1 → X_1; then

$$p_1\varphi = p_1(\tilde{p}_2\tilde{p}_1) = (p_1\tilde{p}_2)\tilde{p}_1 = p_2\tilde{p}_1 = p_1.$$

Furthermore $\varphi(x_1) = x_1$ and so by Corollary 17.5 the map φ is the identity map, i.e. $\tilde{p}_2\tilde{p}_1 = 1$. By reversing the roles of X_1 and X_2 we see that $\tilde{p}_1\tilde{p}_2 = 1$. Thus $\tilde{p}_1$ and $\tilde{p}_2$ are homeomorphisms and the theorem follows by taking h = $\tilde{p}_1$.

     A special case of Corollary 21.4 is when X_1 and X_2 are simply connected.

21.5   **Corollary**

     Let p_1: X_1 → X, p_2: X_2 → X be two coverings with X_1 and X_2 both simply connected and locally path connected, then there is a homeomorphism h: X_1 → X_2 such that $p_2 h = p_1$.

     There is a converse to Corollary 21.4 which follows easily from Theorem 15.9.

21.6   **Corollary**

     Let p_1: X_1 → X, p_2: X_2 → X be coverings with X_1, X_2 connected and locally path connected. Let x_1, x_2, x_0 be base points with $p_1(x_1) = p_2(x_2) = x_0$. If there is a homeomorphism h: X_1 → X_2 with $p_2 h = p_1$ and $h(x_1) = x_2$ then

$$p_{1_*} \pi(X_1, x_1) = p_{2_*} \pi(X_2, x_2).$$

     We say that two coverings p_1: X_1 → X, p_2: X_2 → X are *equivalent* if there is a homeomorphism h: X_1 → X_2 such that $p_2 h = p_1$. Note that the base points of X_1 and X_2 are not necessarily preserved by the homeomorphism h. Corollaries 21.4 and 21.6 generalize to give:

21.7 Theorem

Let $p_1: X_1 \to X$, $p_2: X_2 \to X$ be coverings with X_1, X_2 both connected and locally path connected. Let x_1, x_2, x_0 be base points with $p_1(x_1) = p_2(x_2) = x_0$. The two coverings are equivalent if and only if the subgroups $p_1{}_* \pi(X_1, x_1)$ and $p_2{}_* (X_2, x_2)$ of $\pi(X, x_0)$ are conjugate.

Proof This follows directly from Corollaries 21.4, 21.6 and Theorem 18.3.

The *group of covering transformations* of a covering $p: \tilde{X} \to X$ is the group of all homeomorphisms $h: \tilde{X} \to \tilde{X}$ such that $ph = p$ (see Exercise 17.9(r)). This group is denoted by $G(\tilde{X}, p, X)$. Clearly $\tilde{X}$ is a $G(\tilde{X}, p, X)$-space.

21.8 Theorem

If $\tilde{X}$ is connected and locally path connected then the action of $G(\tilde{X}, p, X)$ on $\tilde{X}$ is properly discontinuous.

Proof Let x be any point of $\tilde{X}$ and let U be an evenly covered neighbourhood of $p(x)$. Thus $p^{-1}(U)$ is the disjoint union of $\{ V_j, j \in J \}$ with $x \in V_k$ for some k. Let $h \in G(\tilde{X}, p, X)$. If $h(x) = x$ then by Corollary 17.5 the map h is the identity. In other words if $h \neq 1$ then $h(x) \neq x$. Since $ph(x) = p(x)$ it follows that $h(x) \in V_\ell$ for some ℓ; furthermore if $V_\ell = V_k$ then $h(x) = x$. We therefore conclude that if $h \neq 1$ then $x \in V_k$ and $h(x) \in V_\ell$ with $V_k \cap V_\ell = \emptyset$.

We may insist that U is path connected since $\tilde{X}$ (and hence X clearly) is locally path connected. Thus each of the sets V_j, $j \in J$, are path connected. Now $ph(V_k) = U$, so that $h(V_k) \subseteq \underset{j \in J}{\cup} V_j$. But the V_j, $j \in J$, are path connected and $h(x) \in V_\ell$ for some $x \in V_k$ which means that $h(V_k) \subseteq V_\ell$ and hence $V_k \cap h(V_k) = \emptyset$. This proves that the action of $G(\tilde{X}, p, X)$ is properly discontinuous.

Using this result and some previous results leads to the following interesting results.

21.9 Theorem

Let $\tilde{X}$ be connected and locally path connected. If $p_* \pi(\tilde{X}, \tilde{x}_0)$ is a normal subgroup of $\pi(X, x_0)$ then X is homeomorphic to $\tilde{X}/G(\tilde{X}, p, X)$.

Proof Since $p_* \pi(\tilde{X}, \tilde{x}_0)$ is a normal subgroup we see from Theorem 18.3 that

$$p_* \pi(\tilde{X}, \tilde{x}_1) = p_* \pi(\tilde{X}, \tilde{x}_0)$$

for any $\tilde{x}_1 \in p^{-1}(x_0)$. Hence by Corollary 21.4 there is an element of $G(\tilde{X},p,X)$ such that $h(\tilde{x}_0) = \tilde{x}_1$. Thus if $p(\tilde{x}_0) = p(\tilde{x}_1)$ then there is an element $h \in G(\tilde{X},p,X)$ such that $h(\tilde{x}_0) = \tilde{x}_1$. Conversely, it is clear that if $h(\tilde{x}_0) = \tilde{x}_1$ for some $h \in G(\tilde{X},p,X)$ then $p(\tilde{x}_0) = p(\tilde{x}_1)$. Thus the group $G(\tilde{X},p,X)$ identifies points in $\tilde{X}$ in the same way as p does. This shows that X and $\tilde{X}/G(\tilde{X},p,X)$ coincide setwise. They are homeomorphic because each has the natural quotient topology determined by the projections $p\colon \tilde{X} \to X$ and $\pi\colon \tilde{X} \to \tilde{X}/G(\tilde{X},p,X)$.

21.10   Corollary
   Let $\tilde{X}$ be connected and locally path connected. If $p_* \pi(\tilde{X},\tilde{x}_0)$ is a normal subgroup of $\pi(X,x_0)$ then
$$\pi(X,x_0)/p_* \pi(\tilde{X},\tilde{x}_0) \cong G(\tilde{X},p,X).$$

Proof This follows immediately from Theorem 21.9 and Theorem 19.3.

21.11   Corollary
   If $\tilde{X}$ is simply connected and locally path connected then
$$\pi(X,x_0) \cong G(\tilde{X},p,X).$$

21.12   Exercises
   (a) Show that the subspace P of $\mathbb{R}$ defined by
   $$P = \{\ 0, 1/n; n \text{ is a positive integer }\}$$
   is not locally path connected.
   (b) Let $p_1\colon X_1 \to S^1$, $p_2\colon X_2 \to S^1$ be n-fold coverings (n a finite positive integer). Show that they are equivalent.
   (c) Determine all covering spaces of (i) S^1, (ii) the torus $S^1 \times S^1$ and (iii) a space X which is simply connected and locally path connected.
   (d) Let $p_1\colon X_1 \to X$, $p_2\colon X_2 \to X$ be covering maps with X connected and locally path connected. (i) Prove that if there is a continuous surjection $f\colon X_1 \to X_2$ then $f\colon X_1 \to X_2$ is a covering map. (ii) Prove that if X_2 is path connected and if there is a continuous map $f\colon X_1 \to X_2$ then $f\colon X_1 \to X_2$ is a covering map.
   (e) Suppose that $p_1\colon X_1 \to X$ and $p_2\colon X_2 \to X$ are coverings in which X_1 is simply connected and locally path connected while X_2 is connected and locally path connected. Prove that there is a continuous map $p\colon X_1 \to X_2$ which is a covering map.
   (f) Suppose that X is a connected G-space for which the action of G on X is properly discontinuous. Prove that the group of covering

transformations of p: $X \to X/G$ is G.

(g) Let p: $\widetilde{X} \to X$ be a covering with $\widetilde{X}$ connected and locally path connected. Prove that $G(\widetilde{X}, p, X)$ acts on $p^{-1}(x_0)$. Furthermore prove that $G(\widetilde{X}, p, X)$ acts transitively on $p^{-1}(x_0)$ if and only if $p_*\pi(\widetilde{X}, \widetilde{x}_0)$ is a normal subgroup of $\pi(X, x_0)$. (See Exercise 18.4(*b*) for the definition of a transitive action.)

22

More on covering spaces: existence theorems

In the last chapter we showed that a covering p: $\widetilde{X} \to X$ is determined, up to equivalence, by the conjugacy class of the subgroup $p_* \pi(\widetilde{X},\widetilde{x}_0)$ of $\pi(X,x_0)$. It is reasonable to ask whether for a given conjugacy class of subgroups of $\pi(X,x_0)$ there exists a covering p: $\widetilde{X} \to X$ which belongs to the given conjugacy class. The answer, as we shall show later on, is yes, provided that we put some extra conditions on X (in addition to X being connected and locally path connected). There is always a covering corresponding to the conjugacy class of the entire fundamental group of X, namely 1: $X \to X$. This, however, is of little interest. At the other extreme, the covering that corresponds to the conjugacy class of the trivial subgroup is very interesting. This covering, if it exists for a given X, is called the *universal covering* of X. Thus the universal covering of X is a covering p: $\widetilde{X} \to X$ for which $\widetilde{X}$ is simply connected. Shortly we shall give a necessary and sufficient condition on X to ensure that a universal covering for X exists.

Suppose that p: $\widetilde{X} \to X$ is a universal covering of X. If x is any point of X and $\widetilde{x} \in p^{-1}(x)$ then there is an evenly covered neighbourhood U of x with $p^{-1}(U)$ being the disjoint union of $\{ V_j; j \in J \}$ and $\widetilde{x} \in V_k$ for some k. Denote V_k by V. The diagram

$$
\begin{array}{ccc}
V & \longrightarrow & \widetilde{X} \\
{\scriptstyle p|V} \downarrow & {\scriptstyle i} & \downarrow {\scriptstyle p} \\
U & \longrightarrow & X
\end{array}
$$

leads to a commutative diagram of fundamental groups

$$
\begin{array}{ccc}
\pi(V,\widetilde{x}) & \longrightarrow & \pi(\widetilde{X},\widetilde{x}) \\
{\scriptstyle (p|V)_*} \downarrow & {\scriptstyle i_*} & \downarrow {\scriptstyle p_*} \\
\pi(U,x) & \longrightarrow & \pi(X,x)
\end{array}
$$

The map p|V: $V \to U$ is a homeomorphism and so $(p|V)_*$ is an isomorphism. Since p: $\widetilde{X} \to X$ is a universal covering, the group $\pi(\widetilde{X},\widetilde{x})$ is trivial and so the map i_* must be the trivial homomorphism, i.e. $i_*(\alpha) = [\epsilon_x]$ for all $\alpha \in \pi(U,x)$. This shows that if p: $\tilde{X} \to X$ is a universal covering of X then every point $x \in X$ has a neighbourhood U such that the homomorphism $\pi(U,x) \to \pi(X,x)$ is trivial. A space X with this property is said to be 'semilocally simply connected'. Thus a space is *semilocally simply connected* if and only if for every x in X there is a neighbourhood U of x such that any closed path in U, based at x, is equivalent in X to the constant path ϵ_x. Note that in the process of making a closed path in U equivalent to ϵ_x we may go outside U. Note also that if U is a neighbourhood of x such that every closed path in U, based at x, is equivalent in X to the constant path then every neighbourhood U' of x such that $U' \subseteq U$ also has the property that every closed path in U', based at x, is equivalent in X to the constant path. The most interesting spaces are semilocally simply connected (see the exercises) and we have to think quite hard to obtain an example of a space that is connected and locally path connected but not semilocally simply connected. Such an example is the subspace X of $\mathbb{R}^2$ given by

$$X = \bigcup_{n>0} C_n$$

where C_n is the circle with centre $(1/n, 0) \in \mathbb{R}^2$ and radius $1/n$. The point $(0,0) \in X$ fails to satisfy the condition required for X to be semilocally simply connected. Thus this space fails to have a universal covering.

The above necessary condition on X for the existence of a universal covering is in fact also sufficient.

22.1   Theorem

Let X be a connected and locally path connected space. Then X has a universal covering p: $\tilde{X} \to X$ if and only if X is semilocally simply connected.

Proof We first construct a space $\tilde{X}$ and a map p: $\tilde{X} \to X$, then we show that these have the required properties. Let $x_0 \in X$ be a base point and let $\tilde{X}$ be the set of equivalence classes of paths that begin at x_0 (see Definition 14.1). Thus

$$\tilde{X} = \{ [\alpha] ; \alpha: I \to X, \alpha(0) = x_0, [\alpha] = [\beta] \Leftrightarrow \alpha \sim \beta \} .$$

Now define p: $\tilde{X} \to X$ by $p([\alpha]) = \alpha(1)$.

We have to put a topology on $\tilde{X}$. Let U be an open set in X and let α: $I \to X$ be a path that begins at x_0 and ends at some point, say x_1, in U. Define $[U,\alpha]$ by

$$[U,\alpha] = \{ [\alpha * \beta] ; \beta: I \to X, \beta(0) = \alpha(1), \beta(I) \subseteq U \} .$$

In other words $[U,\alpha]$ consists of the equivalence classes of paths $\alpha * \beta$ for which β begins at $\alpha(1)$ and β lies entirely within U. We use these sets to define a topology $\mathcal{U}$ for $\widetilde{X}$ as follows: $\mathcal{U}$ consists of $\emptyset$, $\widetilde{X}$ and arbitrary unions of subsets of $\widetilde{X}$ of the form $[U,\alpha]$. To check that $\mathcal{U}$ is a topology for $\widetilde{X}$ we need only check that the intersection of two members of $\mathcal{U}$ is still a member of $\mathcal{U}$ (the other conditions for a topology are trivially satisfied). First we show that if $[\gamma] \in [U,\alpha]$ then $[U,\gamma] = [U,\alpha]$. We see this as follows. Since $[\gamma] \in [U,\alpha]$ there is some path β lying in U such that $[\gamma] = [\alpha * \beta]$. If δ is any path in U beginning at $\beta(1)$ then

$$[\gamma * \delta] = [(\alpha * \beta) * \delta] = [\alpha * (\beta * \delta)]$$

which shows that $[U,\gamma] \subseteq [U,\alpha]$. The same argument shows that $[U,\alpha] \subseteq [U,\gamma]$ and so $[U,\gamma] = [U,\alpha]$. Now consider $[U,\alpha] \cap [U',\alpha']$, where $[U,\alpha]$ and $[U',\alpha']$ belong to $\mathcal{U}$. If $\beta \in [U,\alpha] \cap [U',\alpha']$ then $[U,\beta] = [U,\alpha]$ and $[U',\beta] = [U',\alpha']$. Immediately we have

$$[U \cap U', \beta] \subseteq [U,\alpha] \cap [U',\alpha']$$

and hence $[U,\alpha] \cap [U',\alpha']$ is the union of

$$\{ [U \cap U', \beta] ; \beta \in [U,\alpha] \cap [U',\alpha] \}$$

which shows that $[U,\alpha] \cap [U',\alpha'] \in \mathcal{U}$. We leave it for the reader to check that the intersection of *any* two elements of $\mathcal{U}$ belongs to $\mathcal{U}$ (it is easy). Thus $\mathcal{U}$ is a topology for $\widetilde{X}$.

We now check that the map p: $\widetilde{X} \to X$ is continuous. Let U be an open subset of X. If $p^{-1}(U)$ is empty then we are finished. Suppose $[\alpha] \in p^{-1}(U)$; then by definition $[U,\alpha]$ is an open set of $\widetilde{X}$ and

$$\begin{aligned}
p([U,\alpha]) &= \{ (\alpha * \beta)(1); [\alpha * \beta] \in [U,\alpha] \} \\
&= \{ \beta(1); [\alpha * \beta] \in [U,\alpha] \} \\
&\subseteq U,
\end{aligned}$$

since by definition of $[U,\alpha]$ the paths β lie in U. We therefore have

$$p^{-1}(U) = \bigcup_{[\alpha] \in p^{-1}(U)} [U,\alpha]$$

which is an open set in $\widetilde{X}$, and so p is continuous.

Next we check that p: $\widetilde{X} \to X$ is surjective. This is easy because if $x \in X$ then there is a path in X beginning at x_0 and ending at x (the space X is path connected). Clearly $[\alpha] \in \widetilde{X}$ and $p([\alpha]) = x$.

To show that p: $\widetilde{X} \to X$ is a covering it remains for us to show that each point of X has a neighbourhood which is evenly covered. Let $x \in X$ and let V be an open neighbourhood of x which is path connected and for which every closed path based at x is equivalent in X to the constant path ϵ_x. We have

$$p^{-1}(V) = \bigcup_{[\alpha] \in p^{-1}(V)} [V,\alpha].$$

If $[V,\alpha] \cap [V,\beta] \neq \emptyset$ then there is an element $[\gamma] \in [V,\alpha] \cap [V,\beta]$ and hence $[V,\gamma] = [V,\alpha]$ and $[V,\gamma] = [V,\beta]$ so that $[V,\alpha] = [V,\beta]$. This shows that $p^{-1}(V)$ is a disjoint union of open sets. We must show that p maps each set homeomorphically onto V. The map

$$p_\alpha = p | [V,\alpha] : [V,\alpha] \to V$$

is obviously continuous since $p_\alpha^{-1}(V) = [V,\alpha]$. If $x \in V$, let β be a path in V from $\alpha(1)$ to x, then $[\alpha * \beta] \in [V,\alpha]$ and $p([\alpha * \beta]) = x$. This shows that p_α is surjective.

To prove that p_α is injective suppose that $p_\alpha([\alpha * \beta]) = p_\alpha([\alpha * \gamma])$ for some two elements $[\alpha * \beta]$, $[\alpha * \gamma] \in [V,\alpha]$. Then β and γ have the same end points. The path $\beta * \bar\gamma$ is a closed path in V and so, by our choice of V, is equivalent in X to the constant path ϵ_x. In particular $\beta \sim \gamma$ and so $[\alpha * \beta] = [\alpha * \gamma]$ proving that p_α is injective.

To complete the proof that p_α is a homeomorphism we need to check that p_α^{-1} is continuous, or equivalently that p_α is an open map. Let $[W,\beta]$ be an open subset of $[V,\alpha]$. Then $N = p([W,\beta])$ is the set of points in W which can be joined by a path in W to $\beta(1)$. For each $y \in N$ there is an open path connected subset W_y of W containing y. Since $y \in N$ and W_y is path connected it follows that $W_y \subseteq N$ and so $N = \bigcup_{y \in N} W_y$. Thus $p([W,\beta]) = N$ is open and so p_α is a homeomorphism, which completes the proof of the statement that $p: \widetilde{X} \to X$ is a covering.

To complete the proof of Theorem 22.1 we need to show that $\widetilde{X}$ is simply connected. First we show that $\widetilde{X}$ is path connected. Let $\widetilde{x}_0 = [\epsilon]$ be the class of the constant path ϵ at x_0 and let $[\alpha]$ be any element of $\widetilde{X}$. Define $\widetilde{\alpha}: I \to \widetilde{X}$ by

$$\widetilde{\alpha}(s) = [\alpha_s] \quad s \in I$$

where $\alpha_s(t) = \alpha(st)$, $t \in I$. Then $\widetilde{\alpha}$ is a path in $\widetilde{X}$ from $\widetilde{x}_0$ to $[\alpha]$. This proves that $\widetilde{X}$ is path connected. (Note that $\widetilde{\alpha}$ is a lift of α.)

Let β be a closed path in $\widetilde{X}$ based at $\widetilde{x}_0$. By uniqueness of liftings $\beta = \widetilde{p\beta}$ and so

$$[p\beta] = [p(\widetilde{p\beta})] = [\widetilde{p\beta}(1)] = \widetilde{x}_0 = [\epsilon].$$

Thus $\beta = \widetilde{p\beta}$ is equivalent to the constant path in $\widetilde{X}$ and so $\widetilde{X}$ is simply connected and the proof of Theorem 22.1 is complete.

22.2   Corollary

Suppose that X is a connected, locally path connected and semi-locally simply connected space. If H is a subgroup of $\pi(X,x_0)$ then there exists a covering p_H: $X_H \to X$, unique up to equivalence, such that $H = p_{H*} \pi(X_H,x_H)$. Thus, in particular, for any conjugacy class of subgroups of $\pi(X,x_0)$ there is a covering space p': $X' \to X$ such that $p'_* \pi(X',x')$ belongs to that conjugacy class.

Proof Let p: $\widetilde{X} \to X$ be the universal covering space of X and let $G(\widetilde{X},p,X)$ be the group of covering transformations. Since $G(\widetilde{X},p,X) \cong \pi(X,x_0)$ we let H' be the subgroup corresponding to H under this isomorphism. We then take $X_H = \widetilde{X}/H'$ and let p_H be the map induced by p. The details are left for the reader.

The conditions on X in Corollary 22.2, ensuring the existence of X_H, can be weakened; see Exercise 22.3(e).

22.3   Exercises

(a) Prove that a simply connected space is semilocally simply connected.

(b) Prove that a connected n-manifold is semilocally simply connected. Prove that a connected n-manifold M has a universal covering p: $\widetilde{M} \to M$ in which $\widetilde{M}$ is also an n-manifold.

(c) Prove that $\pi(S^n,x_0)$ is trivial for $n > 1$ in the following way. Let p: $\widetilde{S}^n \to S^n$ be the universal covering. Define f: $D^n \to S^n$ to send ∂D^n to x_0. Show that f lifts to f': $D^n \to \widetilde{S}^n$ and prove that this yields a continuous map f'': $S^n \to \widetilde{S}^n$ such that $pf'' = 1$. Finally apply the fundamental group to the sequence $S^n \xrightarrow{f''} \widetilde{S}^n \xrightarrow{p} S^n$ to deduce that $\pi(S^n,x_0)$ is trivial.

(d) Let X be a connected, locally path connected and semilocally simply connected space. Let H be a subgroup of $\pi(X,x_0)$. Let $\mathscr{P}$ be the set of paths in X that begin at x_0. Define a relation $\sim_H$ on $\mathscr{P}$ by saying that

$$\alpha \sim_H \beta \Leftrightarrow \alpha(1) = \beta(1) \text{ and } [\alpha * \bar{\beta}] \in H.$$

Prove that $\sim_H$ is an equivalence relation on $\mathscr{P}$. Denote the equivalence class of α by $[\alpha]_H$. Define X_H to be $\mathscr{P}/\sim_H$ and let p_H: $X_H \to X$ be defined by $p_H([\alpha]_H) = \alpha(1)$. Let U be an open set in X and let α: $I \to X$ be a path in X beginning at x_0 and ending in U. Define $[U,\alpha]_H$ to be

$$[U,\alpha]_H = \{ [\alpha * \beta]_H; \beta: I \to X, \beta(0) = \alpha(1), \beta(I) \subseteq U \}$$

Show that the collection $\mathcal{U}_H$ consisting of $\emptyset$, X_H and arbitrary unions of sets of the form $[U,\alpha]_H$ forms a topology for X_H. Prove finally that $p_H: X_H \to X$ is a covering and that

$$p_{H*}\, \pi(X_H,x_H) = H \subseteq \pi(X,x_0).$$

(e) Let X be a connected and locally path connected space. Let H be a subgroup of $\pi(X,x_0)$. Prove that there exists a covering $p_H: X_H \to X$ such that $p_{H*}\, \pi(X_H,x_H) = H$ if and only if for every point x of X there is a neighbourhood U of x such that any closed path in U based at x is equivalent in X to some element of $H \subseteq \pi(X,x_0)$. (Hint: Modify the proof of (d) above.)

23

The Seifert-Van Kampen theorem: I Generators

The theorem that we are about to discuss gives a quite general method for calculating fundamental groups. It was first proved in the early 1930s independently by H. Seifert and E. Van Kampen. The theorem is frequently called the Van Kampen theorem and sometimes the theorem of Seifert (usually depending upon whether you are English speaking or German speaking).

Suppose that we are given a space X which is the union of two subspaces U_1, U_2 which are both open and path connected. Suppose furthermore that $U_1 \cap U_2$ is non-empty and path connected. The Seifert-Van Kampen theorem gives a way of calculating the fundamental group of X provided that we know the fundamental groups of U_1, U_2 and $U_1 \cap U_2$. (A special case of this theorem appears in Exercise 15.16(c).)

Let $x_0 \in U_1 \cap U_2$ and let $\psi_j \colon U_j \to X$ for $j = 1,2$ denote the inclusion maps. Then roughly speaking the Seifert-Van Kampen theorem tells us

(i) (The 'generators' of $\pi(X, x_0)$.) If $\alpha \in \pi(X, x_0)$ then

$$\alpha = \prod_{k=1}^{n} \psi_{\lambda(k)*} \alpha_k$$

where $a_k \in \pi(U_{\lambda(k)}, x_0)$, $\lambda(k) = 1$ or 2.

(ii) (The 'relators' or 'relations' of $\pi(X, x_0)$.) Let

$$\alpha = \prod_{k=1}^{n} \psi_{\lambda(k)*} \alpha_k$$

be an element of $\pi(X, x_0)$. Then $\alpha = 1$ if and only if α can be reduced to 1 by a finite sequence of operations each of which inserts or deletes an expression from a certain list. This list depends on $\pi(U_1 \cap U_2, x_0)$, $\pi(U_1, x_0)$ and $\pi(U_2, x_0)$.

The information in (i) and (ii) is called a presentation of the group $\pi(X, x_0)$. Thus in order to state the Seifert-Van Kampen theorem precisely we need to specify in detail what a presentation of a group is. It is in fact an elegant way of expressing the group in terms of generators and relators (or relations). Before we say more about this, however, we must introduce some notation.

Consider a set S and think of the elements of S as being non-commutative symbols. Using these symbols we form *words*; these are just expressions of the form

$$W = x_1{}^{\epsilon(1)}x_2{}^{\epsilon(2)} \ldots x_k{}^{\epsilon(k)}$$

where $x_i \in S$, repetitions being allowed, and $\epsilon(i) = \pm 1$. (In other words from the set $S = \{ y_j; j \in J \}$ we get an 'alphabet' $\{ y_j{}^1, y_j^{-1}; j \in J \}$ which is used to form 'words'.) It is convenient to have the *empty word* in which no symbols appear. A word is said to be *reduced* if it does not contain x^1 followed by x^{-1} or vice versa x^{-1} followed by x^1 for some $x \in S$. Thus $x_1{}^1x_1{}^1x_1{}^1$ is a reduced word but $x_1{}^1x_1{}^{-1}x_1{}^1$ is not. Every word can be reduced to a reduced word simply by deleting pairs like x^1x^{-1} or $x^{-1}x^1$ (where $x \in S$) whenever they appear in a word. For example $x_1{}^{-1}x_2{}^1x_2{}^{-1}x_1{}^1x_3{}^1$ reduces to $x_1{}^{-1}x_1{}^1x_3{}^1$ which reduces to $x_3{}^1$.

Using juxtaposition of reduced words as a law of composition and reducing the resulting word, if necessary, it turns out that the set G of reduced words in the symbols of S forms a group. The empty word acts as the identity, and the inverse of the word $W = x_1{}^{\epsilon(1)}x_2{}^{\epsilon(2)} \ldots x_k{}^{\epsilon(k)}$ is given by

$$W^{-1} = x_k{}^{-\epsilon(k)} x_{k-1}{}^{-\epsilon(k^{-1})} \ldots x_2{}^{-\epsilon(2)} x_1{}^{-\epsilon(1)}$$

We leave it for the reader to verify the group axioms. This group is called the *free group generated by* S. Of course the actual symbols in S themselves do not matter, so that if S' is another set bijective to S then the resulting groups generated by S and S' are isomorphic. When S is finite with n elements we call the free group generated by S the *free group on n generators*.

Note that the free group on 1 generator $\{ x \}$ consists of the following elements

$$1, x^1, x^{-1}, x^1x^1, x^{-1}x^{-1}, x^1x^1x^1, x^{-1}x^{-1}x^{-1}, \ldots$$

and it is not difficult to see that it is isomorphic to the group of integers Z. We often abbreviate x^1 by x, x^1x^1 by $x^2, x^{-1}x^{-1}$ by x^{-2} etc. Note also that the free group on n generators for $n > 1$ is a non-abelian infinite group.

It is convenient to view free groups in a slightly different way by considering equivalence classes of words under a suitable equivalence relation. Consider the following operations on words:

(i) insert xx^{-1} or $x^{-1}x$ in a word, where $x \in S$,
(ii) delete xx^{-1} or $x^{-1}x$ in a word, where $x \in S$

(by inserting xx^{-1} in a word W we mean write W as W_1W_2 and then insert

xx^{-1} thus: $W_1 xx^{-1} W_2$; here of course W_1 or W_2 may be empty). We say that two words W, W' are equivalent if and only if W' can be obtained from W by a finite number of operations of type (i) and (ii). This is clearly an equivalence relation; furthermore it is obvious that any word is equivalent to its reduced form. The set of equivalence classes of words in S, with juxta-position as a law of composition, forms the free group generated by S. For brevity we usually denote the equivalence class containing the word W by W itself; this should not lead to any confusion.

Suppose now that R is a set of words in S. We can consider the following additional operations on words in S:

(iii) insert r or r^{-1} in a word, where $r \in R$,

(iv) delete r or r^{-1} in a word, where $r \in R$.

We now say that two words W, W' are equivalent if and only if W' can be obtained from W by a finite number of operations of types (i), (ii), (iii) and (iv). The reader can readily verify that this is an equivalence relation and that the set of equivalence classes forms a group, with juxtaposition as a law of composition. This group is said to be the *group with presentation* (S;R) and is denoted by $\langle S;R \rangle$. As above we denote the equivalence class containing the word W by W itself, again this should not cause any confusion. The set S is called the *generators* and the set R is called the set of *relators*. We shall give three (simple) examples. First, the group with presentation $(S;\emptyset)$ is just the free group generated by S. The second example is $\langle \{x\} ; \{ x^n \} \rangle$, where n is some fixed positive integer. This group consists of the words

$$1, x, x^2, ..., x^{n-1}$$

and is easily seen to be isomorphic to the cyclic group $\mathbb{Z}_n$. For the third example consider $\langle \{x,y\} ; \{ xyx^{-1}y^{-1} \} \rangle$. We see that $xy = yx$, $(xy = (xyx^{-1}y^{-1})^{-1}xy = yxy^{-1}x^{-1}xy = yxy^{-1}y = yx$ by operations (iii) and (ii)). It is then not difficult to see that $x^a y^b = y^b x^a$ for all integers a,b and so any word $g = x^{a(1)}y^{b(1)}x^{a(2)} ... x^{a(k)}y^{b(k)}$ can be rewritten as $g = x^a y^b$ where $a = \Sigma_{i=1}^k a(i)$ and $b = \Sigma_{i=1}^k b(i)$. Thus the group $\langle \{ x,y \} ; \{ xyx^{-1}y^{-1} \} \rangle$ is isomorphic to $\mathbb{Z} \times \mathbb{Z}$.

If α is a word in S and $\alpha = 1$ in $\langle S;R \rangle$ then α itself does not necessarily belong to R, of course. However, α can be reduced to the empty word by a finite sequence of operations of type (i), (ii), (iii) and (iv). In such a case we say that α is a *consequence* of the relators R. For example $x^a y^b x^{-a} y^{-b}$ is a consequence of the relator $xyx^{-1}y^{-1}$.

Different presentations may give rise to isomorphic groups; for example, the group $\langle \{x,y\} ; \{ y \} \rangle$ is isomorphic to the group $\langle \{x\}, \emptyset \rangle$. Similarly the group $\langle \{a,b\} ; \{ baba^{-1} \} \rangle$ is isomorphic to the group $\langle \{a,c\} ;$

$\{a^2c^2\}$. To see this define

$$f: \langle\{a,b\}; \{baba^{-1}\}\rangle \to \langle\{a,c\}; \{a^2c^2\}\rangle$$

on the generators by $f(a) = a$, $f(b) = ca$ and in general by

$$f(x_1{}^{\epsilon(1)}x_2{}^{\epsilon(2)}...x_n{}^{\epsilon(n)}) = f(x_1)^{\epsilon(1)}f(x_2)^{\epsilon(2)}...f(x_n)^{\epsilon(n)}.$$

Since

$$f(baba^{-1}) = caacaa^{-1} = ca^2c = c(a^2c^2)c^{-1} = cc^{-1} = 1,$$

we get a well-defined function which is easily seen to be a homomorphism. If we define

$$g: \langle\{a,c\}; \{a^2c^2\}\rangle \to \langle\{a,b\}; \{baba^{-1}\}\rangle$$

on the generators by $g(a) = a$, $g(c) = ba^{-1}$ it is easy to check that g is well defined, is a homomorphism and $fg = 1$, $gf = 1$ so that f and g are group isomorphisms.

The problem of determining whether or not two presentations determine isomorphic groups is in general extremely difficult. Even if we are told that two groups are isomorphic it may be difficult to see why. For example the group $\langle\{x,y\}; \{xy^2x^{-1}y^{-3}, yx^2y^{-1}x^{-3}\}\rangle$ is in fact isomorphic to the trivial group 1, but this is extremely hard to prove. (The reader should nevertheless try proving that the group is indeed trivial.) Despite this pessimistic tone there are various tricks which will enable us to tell if two groups are different. These tricks will be produced when needed.

Given any group G then we say that G has a presentation (S;R) if G is isomorphic to $\langle S;R\rangle$. Every group has a presentation $(S_G;R_G)$ where

$$S_G = \{g \in G\}, \quad R_G = \{(xy)^1 y^{-1} x^{-1}; x,y \in G\};$$

here $(xy)^1$ means the symbol representating $xy \in G$. Showing that G is isomorphic to $\langle S_G;R_G\rangle$ is left as an exercise for the reader.

It is more convenient, sometimes, to write the relators R of the group $\langle S;R\rangle$ as a set of *relations*. By this we mean that the set $\{r; r \in R\}$ is rewritten as $\{r = 1; r \in R\}$. Furthermore, if r is the product of two words uv then we may replace $r = 1$ by $u = v^{-1}$. Thus, for example, we may write $\langle\{A,B\}; \{ABA^{-1}B^{-1}\}\rangle$ as $\langle\{A,B\}; \{ABA^{-1}B^{-1} = 1\}\rangle$ or as $\langle\{A,B\}; \{AB = BA\}\rangle$. As another example, the relators R_G above may be written as $\{(xy)^1 = x^1y^1; x, y \in G\}$. This informality of notation should not lead to any confusion.

23.1 Exercises

(a) Show that if G is a group then G is isomorphic to $\langle S_G;R_G\rangle$ where

$$S_G = \{g \in G\}, \quad R_G = \{(xy)^1 y^{-1} x^{-1}; x\, y \in G\}.$$

(b) What is the order of the group $\langle\{A,B\};R\rangle$ where
$$R = \{ A^4, A^2B^{-2}, A^3BA^{-1}B^{-1} \} ?$$

(c) Show that the group $\langle\{A,B\};\{A^4,B^2,ABA^{-1}B^{-1}\}\rangle$ is isomorphic to $\mathbb{Z}_4 \times \mathbb{Z}_2$.

(d) Suppose that G is a group with presentation $(S;R)$. Let AG be the group with presentation $(S;AR)$ where
$$AR = R \cup \{ xyx^{-1}y^{-1} ; x,y \in S \}.$$

Show that AG is an abelian group and that there is an epimorphism $G \to AG$. What is the kernel of this epimorphism?

(e) Using Exercise 19.5(e) show that the fundamental group of the Klein bottle has presentation $(\{a,b\} ; \{abab^{-1}\})$.

(f) Let $G = \langle S;R \rangle$. Let T_i $(i = 1,2,3,4)$ be the following transformations on the pair $(S;R)$, the so-called *Tietze transformations*.

T_1: If r is a word in S and r = 1 is a relation which holds in G then let $S' = S, R' = R \cup \{r\}$.

T_2: If $r \in R$ is such that the relation r = 1 holds in the group $\langle S;R - \{r\}\rangle$ then let $S' = S$, $R' = R - \{r\}$.

T_3: If w is a word in S and x is a symbol not in S, let $S' = S \cup \{x\}$, $R' = R \cup \{wx^{-1}\}$.

T_4: If $x \in S$ and if w is a word in S not involving x or x^{-1} such that $wx^{-1} \in R$ then substitute w for x in every element of $R - \{wx^{-1}\}$ to get R' and let $S' = S - \{x\}$.

Prove that if $(S'';R'')$ results from $(S;R)$ by a finite sequence of Tietze transformations then $\langle S'';R''\rangle$ is isomorphic to $\langle S;R\rangle$.

(The transformations T_1 and T_2 correspond to adding and removing a superfluous relation respectively while T_3 and T_4 correspond to adding and removing a superfluous generator respectively.)

Let us now return to our topological space X which is the union of two open path connected subsets U_1 and U_2 with $U_1 \cap U_2$ non-empty and path connected. Let $\varphi_1, \varphi_2, \psi_1, \psi_2$ denote the various inclusion maps as indicated below

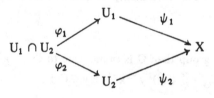

Choose, as a base point, a point $x_0 \in U_1 \cap U_2$. We then have the following commutative diagram of homomorphisms:

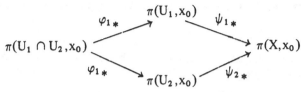

Suppose that the fundamental groups of $U_1 \cap U_2$, U_1 and U_2 are known and that the presentations of these groups are as given:

$$\pi(U_1 \cap U_2, x_0) = \langle S; R \rangle,$$
$$\pi(U_1, x_0) \quad = \langle S_1; R_1 \rangle,$$
$$\pi(U_2, x_0) \quad = \langle S_2; R_2 \rangle.$$

If $s \in S$ then $\varphi_{1*} s \in \pi(U_1, x_0)$ and $\varphi_{2*} s \in \pi(U_2, x_0)$, so that we can express these elements as words in S_1, S_2 respectively. Let '$\varphi_{1*}s$', '$\varphi_{2*}s$' be representations of $\varphi_{1*}s$, $\varphi_{2*}s$ as words in the generators S_1, S_2 respectively.

23.2   Definition
Let R_S denote the following set of words in $S_1 \cup S_2$:

$$('\varphi_{1*}s') ('\varphi_{2*}s')^{-1} \qquad s \in S.$$

We shall think of R_S as a set of relators. As a set of relations R_S is

$$\{ '\varphi_{1*}s' = '\varphi_{2*}s'; s \in S \}.$$

We can now state the theorem of H. Seifert and E. Van Kampen.

23.3   The Seifert–Van Kampen theorem
$\pi(X, x_0)$ is isomorphic to the group defined by the generators $S_1 \cup S_2$ and the relations $R_1 \cup R_2 \cup R_S$.

Note that the relations R of $\pi(U_1 \cap U_2, x_0)$ are not required.

Loosely speaking $\pi(X, x_0)$ is the smallest group generated by $\pi(U_1, x_0)$ and $\pi(U_2, x_0)$ for which $\varphi_{1*}s = \varphi_{2*}s$, $s \in \pi(U_1 \cap U_2, x_0)$.

The proof of the theorem will be divided into essentially two parts. The first part will be concerned with generators and will be proved in this chapter. The second part will be concerned with relations and will be proved in the next chapter. We now prove a result that will be useful (in fact it is just Exercise 7.13(g)); we could, as in Chapter 16, avoid using this result.

23.4   Theorem
Let X be a compact topological space arising from some metric

space with metric d. Given an open cover $\{\ U_j;\ j \in J\ \}$ then there exists a real number $\delta > 0$ (called the *Lebesgue number* of $\{\ U_j;\ j \in J\ \}$) such that any subset of diameter less than δ is contained in one of the sets U_j, $j \in J$.

Proof Since X is compact we may assume that J is finite. For $x \in X$ and $j \in J$ let $f_j(x)$ be given by

$$f_j(x) = d(x, X - U_j) = \inf_{y \in X-U_j} d(x,y).$$

Clearly f_j is continuous, as is the function f defined by

$$f(x) = \max_{j \in J} f_j(x).$$

Since $X - U_j$ is closed it follows that $f_j(x) = 0$ if and only if $x \in X - U_j$. Thus $f(x) = 0$ if and only if $x \in X - U_j$ for all $j \in J$. But $\{\ U_j; j \in J\ \}$ is a cover of X and so $f(x) > 0$ for all $x \in X$. That X is compact and f is continuous means that $f(X)$ is a compact subset of $\mathbb{R}$, in fact of $(0, \infty) \subset \mathbb{R}$. Therefore there is a $\delta > 0$ such that $f(x) > \delta$ for all $x \in X$. We claim that any set S of diameter less than δ must belong to some U_k, $k \in J$. To see this, simply take $x \in S$; then $f(x) > \delta$ which means that $f_k(x) > \delta$ for some k which in turn means that $x \in U_k$. But the diameter of S is less than δ and $d(x, X - U_k) > \delta$ for some $x \in S$ so that S itself is in U_k, which proves our assertion.

The first step in the proof of the Seifert-Van Kampen theorem is concerned with generators. Essentially we solve Exercises 14.6(g), (h) and (i).

23.5   Lemma

The group $\pi(X, x_0)$ is generated by

$$\psi_{1*}(\pi(U_1, x_0)) \cup \psi_{2*}(\pi(U_2, x_0)).$$

In other words if $\alpha \in \pi(X, x_0)$ then $\alpha = \Pi\ \psi_{\lambda(k)*}\alpha_k$ where $\alpha_k \in \pi(U_{\lambda(k)}, x_0)$ and $\lambda(k) = 1$ or 2.

Proof Let f be a closed path based at $x_0 \in X$. Let δ be the Lebesgue number of the open cover $\{\ f^{-1}(U_1), f^{-1}(U_2)\ \}$ of I. This means that if $t_0, t_1, t_2, \ldots,$ t_n is a sequence of real numbers with

$$0 = t_0 \le t_1 \le t_2 \le \ldots \le t_n = 1$$

and $t_i - t_{i-1} < \delta$ then $f([t_{i-1}, t_i])$ is contained in U_1 or U_2 for $i = 1, 2, \ldots, n$. We may assume that $f(t_i) \in U_1 \cap U_2$. (If $f(t_i) \notin U_1 \cap U_2$ then $[t_{i-1}, t_i]$ and $[t_i, t_{i+1}]$ are either both in U_1 or both in U_2 so that we can combine these two intervals into $[t_{i-1}, t_{i+1}]$ with $f([t_{i-1}, t_{i+1}])$ in U_1 or U_2. Now relabel and continue this process; see Figure 23.1.)

Let paths $f_i : I \to X$, for $i = 1,2,...,n$ be defined by

$$f_i(t) = f((1 - t)t_{i-1} + tt_i).$$

Notice that f_i is a path that is either in U_1 or in U_2, beginning at $f(t_{i-1})$ and ending at $f(t_i)$. We claim that

$$[f] = [f_1] [f_2] ... [f_n].$$

This was in fact Exercise 14.6(i), but we give a proof for completeness.

Figure 23.1

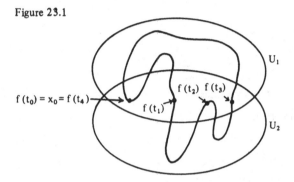

23.6   Lemma

Let $f : I \to X$ be a path and let

$$0 = t_0 \leq t_1 \leq t_2 \leq ... \leq t_n = 1.$$

If $f_j : I \to X$, for $j = 1,2,...,n$ is defined by

$$f_j(t) = f((1 - t)t_{j-1} + tt_j)$$

then

$$[f] = [f_1] [f_2] ... [f_n].$$

Proof The proof is by induction on n. Suppose first that $n = 2$, then $0 = t_0 \leq t_1 \leq t_2 = 1$ and

$$(f_1 * f_2)(t) = \begin{cases} f_1(2t) & 0 \leq t \leq \tfrac{1}{2}, \\ f_2(2t-1) & \tfrac{1}{2} \leq t \leq 1 \end{cases}$$

$$= \begin{cases} f(2tt_1) & 0 \leq t \leq \tfrac{1}{2}, \\ f((1-(2t-1))t_1 + 2t - 1) & \tfrac{1}{2} \leq t \leq 1. \end{cases}$$

We can see that $f_1 * f_2 \sim f$ simply by using the homotopy $F: I \times I \to X$ given by

$$F(t,s) = \begin{cases} f((1-s)2tt_1 + st) & 0 \leq t \leq \tfrac{1}{2}, \\[2mm] f((1-s)(t_1 + (2t-1)(1-t_1)) + st) & \tfrac{1}{2} \leq t \leq 1. \end{cases}$$

Suppose now that $n > 2$ and that the result holds for smaller integers. We have

$$0 = t_0 \leq t_1 \leq \ldots \leq t_n = 1.$$

Since $0 = t_0 \leq t_{n-1} \leq t_n = 1$ we can apply the above result to get

$$f \sim g * f_n$$

where $g(t) = f(tt_{n-1})$. Now

$$0 = \frac{t_0}{t_{n-1}} \leq \frac{t_1}{t_{n-1}} \leq \ldots \frac{t_{n-2}}{t_{n-1}} \leq \frac{t_{n-1}}{t_{n-1}} = 1,$$

so that by the inductive hypothesis

$$[g] = [g_1] [g_2] \cdots [g_{n-1}]$$

where

$$\begin{aligned} g_i(t) &= g((1-t)t_{i-1}/t_{n-1} + tt_1/t_{n-1}) \\ &= f((1-t)t_{i-1} + tt_i) \\ &= f_i(t). \end{aligned}$$

Thus $[f] = [f_1] [f_2] \cdots [f_n]$, which completes the proof.

An alternative proof would be a direct one as follows.

$$((\ldots(((f_1 * f_2) * f_3) * \ldots) * f_n)(t) = \begin{cases} f_1(2^{n-1}t) & 0 \leq t \leq (\tfrac{1}{2})^{n-1}, \\ f_2(2^{n-1}t - 1) & (\tfrac{1}{2})^{n-1} \leq t \leq (\tfrac{1}{2})^{n-2}, \\ \quad \vdots & \quad \vdots \\ f_k(2^{n-k+1}t - 1) & (\tfrac{1}{2})^{n-k+1} \leq t \leq (\tfrac{1}{2})^{n-k}, \\ \quad \vdots & \quad \vdots \\ f_n(2t - 1) & \tfrac{1}{2} \leq t \leq 1 \end{cases}$$

$$= \begin{cases} f(2^{n-1}tt_1) & 0 \leq t \leq (\tfrac{1}{2})^{n-1}, \\ f(t_1 + (2^{n-1}t - 1)t_2 - t_1)) & (\tfrac{1}{2})^{n-1} \leq t \leq (\tfrac{1}{2})^{n-2}, \\ \quad \vdots & \quad \vdots \\ f(t_{k-1} + (2^{n-k+1}t - 1)(t_k - t_{k-1})) & (\tfrac{1}{2})^{n-k+1} \leq t \leq (\tfrac{1}{2})^{n-k}, \\ \quad \vdots & \quad \vdots \\ f(t_{n-1} + (2t - 1)(1 - t_{n-1})) & \tfrac{1}{2} \leq t \leq 1 \end{cases}$$

$$= f(h(t))$$

where h: $I \to I$ is the continuous function given by

$$h(t) = \begin{cases} 2^{n-1} t t_1 & 0 \le t \le (\frac{1}{2})^{n-1}, \\ t_{k-1} + (2^{n-k+1} t - 1)(t_k - t_{k-1}) & (\frac{1}{2})^{n-k+1} \le t \le (\frac{1}{2})^{n-k}, k = 2,3,...,n. \end{cases}$$

Define $F: I \times I \to X$ by

$$F(t,s) = f(sh(t) + (1 - s)t).$$

Clearly F is continuous and so

$f \sim (...((f_1 * f_2) * f_3) *...) * f_n$ which proves the required result.

Returning to the proof of Lemma 23.5, choose, for i = 1,2,...,n-1 paths $q_i: I \to X$ so that $q_i(0) = x_0$, $q_i(1) = f(t_i)$ and so that $q_i(t) \in U_1 \cap U_2$ for all $t \in I$. Also, let q_0 and q_n be given by $q_0(t) = q_n(t) = x_0$. See Figure 23.2.

Figure 23.2

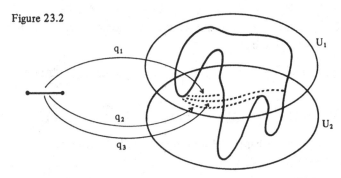

Since $[f] = [f_1] [f_2] ... [f_n]$ we have

$$[f] = [q_0] [f_1] [\bar{q}_1] [q_1] [f_2] [\bar{q}_2] ... [q_{n-1}] [f_n] [\bar{q}_n]$$
$$= [q_0 * f_1 * \bar{q}_1] [q_1 * f_2 * \bar{q}_2] ... [q_{n-1} * f_n * \bar{q}_n]$$

and each of $q_i * (f_{i+1} * \bar{q}_{i+1})$ are closed paths based at x_0 which lie entirely in U_1 or U_2. Hence $[q_i * f_{i+1} * \bar{q}_{i+1}]$ is an element of either $\psi_{1*} \pi(U_1,x_0)$ or $\psi_{2*} \pi(U_2,x_0)$. Thus each element of $\pi(X,x_0)$ may be written as the product of images of elements from $\pi(U_1,x_0)$ and $\pi(U_2,x_0)$ which proves Lemma 23.5.

23.7 Corollary

The group $\pi(X,x_0)$ is generated by the set $\psi_{1*} S_1 \cup \psi_{2*} S_2$ where S_1, S_2 are the generators of $\pi(U_1,x_0)$, $\pi(U_2,x_0)$ respectively.

It is cumbersome to keep writing the ψ_{1*} and ψ_{2*}, so we adopt the convention that we write s in place of ψ_{j*}s for $s \in S_j$, j = 1,2. In other words if f: $I \to U_j$ then we denote the composite $I \xrightarrow{f} U_j \xrightarrow{\psi_j} X$ also by f. In this sense $\pi(X, x_0)$ is generated by $S_1 \cup S_2$ where S_1, S_2 generate $\pi(U_1, x_0), \pi(U_2, x_0)$ respectively.

The next corollary follows immediately from Corollary 23.7.

23.8   Corollary
   If $S_1 = S_2 = \emptyset$ then $\pi(X, x_0)$ is trivial.

A special case of this is:

23.9   Corollary
   If $n \geq 2$ then S^n is simply connected.

This follows from Corollary 23.8 because S^n may be expressed as $U_1 \cup U_2$ where $U_1 = S^n - \{ (1,0,0,0,...,0) \}$, $U_2 = S^n - \{ (-1,0,0,...,0) \}$. Both U_1 and U_2 are simply connected since they are homeomorphic to $\mathbb{R}^n$ with homeomorphisms given by

$$\varphi_1 : U_1 \to \mathbb{R}^n, \quad \varphi_1(x_1,...,x_{n+1}) = \left(\frac{x_2}{1-x_1}, ..., \frac{x_{n+1}}{1-x_1} \right)$$

$$\varphi_2 : U_2 \to \mathbb{R}^n, \quad \varphi_2(x_1,...,x_{n+1}) = \left(\frac{x_2}{1+x_1}, ..., \frac{x_{n+1}}{1+x_1} \right)$$

Using results from Chapter 19 we can now properly deduce:

23.10   Corollary
   The fundamental group of $\mathbb{R}P^n$ is $\mathbb{Z}_2$ and of $L(p,q)$ is $\mathbb{Z}_p$.

23.11   Exercise
   Suppose that $X = \cup_{i=1}^n U_i$ with each U_i, $i = 1,2,...,n$, being open and path connected. Also suppose that $\cap_{i=1}^n U_i$ is non-empty and path connected. Let $x_0 \in \cap_{i=1}^n U_i$. Prove that $\pi(X, x_0)$ is generated by $\cup_{i=1}^n \psi_{i*} \pi(U_i, x_0)$ where $\psi_i : U_i \to X$ denotes the inclusion map.

24

The Seifert-Van Kampen theorem: II Relations

In this chapter we shall complete the proof of the Seifert-Van Kampen theorem (Theorem 23.3). Recall that we are assuming $X = U_1 \cup U_2$ with U_1, U_2 and $U_1 \cap U_2$ being non-empty open path connected subsets of X. Our base point x_0 is in $U_1 \cap U_2 \subseteq X$ and $\pi(U_1 \cap U_2, x_0)$ is generated by S while $\pi(U_j, x_0)$ has the presentation $(S_j; R_j)$ for $j = 1, 2$. Finally R_S is the set of relations '$\varphi_{1*}s$' = '$\varphi_{2*}s$' for $s \in S$. In the previous chapter we showed that $\pi(X, x_0)$ is generated by $S_1 \cup S_2$.

24.1   Lemma
The generators $S_1 \cup S_2$ of $\pi(X, x_0)$ satisfy the relations R_1, R_2 and R_S.

Proof Since

$$\psi_{j*} : \pi(U_j, x_0) \to \pi(X, x_0)$$

is a homomorphism for $j = 1, 2$ any relation satisfied by the elements of S_j in $\pi(U_j, x_0)$ is also satisfied by the elements $\psi_{j*}S_j \subseteq \pi(X, x_0)$. Thus, if we use our convention of suppressing ψ_{j*}, the elements $S_1 \cup S_2$ in $\pi(X, x_0)$ satisfy the relations R_1 and R_2.

If $s \in S \subseteq \pi(U_1 \cap U_2, x_0)$ then

$$\psi_{1*}\varphi_{1*}s = \psi_{2*}\varphi_{2*}s$$

since $\psi_1\varphi_1 = \psi_2\varphi_2$. If a word in S_j represents $\varphi_{j*}s$ then the same word in S_j represents $\psi_{j*}\varphi_{j*}s$ in $\pi(X, x_0)$ so that

$$'\varphi_{1*}s' = '\varphi_{2*}s' \quad s \in S,$$

and so the proof of Lemma 24.1 is finished.

The proof of the Seifert-Van Kampen theorem will be completed when we have shown that the relations mentioned in Lemma 24.1 are the only relations.

24.2   Theorem

   If the elements of $S_1 \cup S_2$ in $\pi(X,x_0)$ satisfy a relation then it is a consequence of the relations R_1, R_2 and R_S.

Proof The proof of this theorem is not difficult but it is quite long and requires a good deal of notation.

   Suppose that $\alpha_1^{\epsilon(1)} \alpha_2^{\epsilon(2)} \ldots \alpha_k^{\epsilon(k)} = 1$ is a relation between the elements of $S_1 \cup S_2 \subseteq \pi(X,x_0)$. Here $\epsilon(i) = \pm 1$ and $\alpha_i \in S_{\lambda(i)}$ for $i = 1, 2, \ldots, k$ where $\lambda(i) = 1$ or 2. For each i, $i = 1, 2, \ldots, k$, choose a closed path f_i in $U_{\lambda(i)}$ based at x_0 such that $[f_i] = \alpha_i^{\epsilon(i)}$. In other words $\alpha_i = [f_i]$ if $\epsilon(i) = 1$ and $\alpha_i = [\bar{f_i}]$ if $\epsilon(i) = -1$. Define a path f: $I \to X$ by

$$f(t) = f_i(kt - i + 1) \text{ for } \frac{(i-1)}{k} \leq t \leq \frac{i}{k}, i = 1, 2, \ldots, k.$$

Notice that

$$f_j(t) = f((1-t)(j-1)/k + tj/k)$$

and since

$$0 = 0/k \leq 1/k \leq \ldots \leq k/k = 1$$

we can use Lemma 23.6 to deduce that

$$[f] = [f_1] [f_2] \ldots [f_k].$$

Since $\alpha_1^{\epsilon(1)} \alpha_2^{\epsilon(2)} \ldots \alpha_k^{\epsilon(k)} = 1$ it follows that $[f] = 1$, i.e. $f \sim \epsilon_{x_0}$. Let F: $I \times I \to X$ be a homotopy between f and ϵ_{x_0}, i.e.

$$F(t, 0) = f(t),$$
$$F(t, 1) = F(0, s) = F(1, s) = x_0.$$

   Now let δ be the Lebesgue number of the open cover { $F^{-1}(U_1)$, $F^{-1}(U_2)$ } of $I \times I$ and choose numbers

$$0 = t_0 < t_1 < t_2 < \ldots < t_m = 1,$$
$$0 = s_0 < s_1 < s_2 < \ldots < s_n = 1,$$

such that

   (i) { $1/k, 2/k, \ldots, (k-1)/k$ } $\subseteq$ { $t_1, t_2, \ldots, t_{m-1}$ },

and

   (ii) $(t_i - t_{i-1})^2 + (s_j - s_{j-1})^2 < \delta^2$ for all i,j.

Clearly such a choice is possible. If $R_{i,j}$ denotes the rectangular region $[t_{i-1}, t_i] \times [s_{j-1}, s_j]$ in $I \times I$ then $F(R_{i,j})$ is contained in either U_1 or U_2 for all i, j.

   For each i,j let $a_{i,j}$: $I \to X$ be a path from x_0 to $F(t_i, s_j)$ which lies in U_1 (or U_2 or $U_1 \cap U_2$) if $F(t_i, s_j)$ lies in U_1 (or U_2 or $U_1 \cap U_2$ respectively).

Such a choice is possible since each of U_1, U_2 and $U_1 \cap U_2$ are path connected. If $F(t_i, s_j) = x_0$ then we insist that $a_{i,j} = \epsilon_{x_0}$. (See Figure 24.1.)

Also, define paths $b_{i,j}$, $c_{i,j}$ as follows

$$b_{i,j}(t) = F((1-t)t_{i-1} + tt_i, s_j),$$
$$c_{i,j}(t) = F(t_i, (1-t)s_{j-1} + ts_j),$$

so that $b_{i,j}$ is a path from $F(t_{i-1}, s_j)$ to $F(t_j, s_j)$ while $c_{i,j}$ is a path from $F(t_i, s_{j-1})$ to $F(t_i, s_j)$; see Figure 24.1. Notice that

$$[f] = [b_{1,0}]\,[b_{2,0}] \cdots [b_{m,1}],$$
$$[\epsilon_{x_0}] = [b_{1,n}]\,[b_{2,n}] \cdots [b_{m,n}]$$

The paths $b_{i,j-1} * c_{i,j}$ and $c_{i-1,j} * b_{i,j}$ are equivalent as paths: intuitively just move the paths within the region $F(R_{i,j})$. An equivalence of paths is given explicitly by the homotopy $H: I \times I \to X$ where

$$H(t,s) = \begin{cases} F((1-s)((1-2t)t_{i-1} + 2tt_i) + st_{i-1}, \\ \qquad (1-s)s_{j-1} + s((1-2t)s_{j-1} + 2ts_j)), \, 0 \le t \le \tfrac{1}{2}, \\[2mm] F((1-s)t_i + s((2-2t)(t_{i-1} + (2t-1)t_i)), \\ \qquad (1-s)((2-2t)s_{j-1} + (2t-1)s_j) + ss_j), \, \tfrac{1}{2} \le t \le 1. \end{cases}$$

Notice that $H(I \times I) \subseteq U_1$, U_2 or $U_1 \cap U_2$ according as $F(R_{i,j}) \subseteq U_1$, U_2 or $U_1 \cap U_2$ respectively.

Figure 24.1

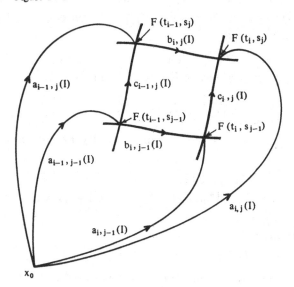

Now define closed paths $f_{i,j}$ and $g_{i,j}$ based at x_0 by

$$f_{i,j} = (a_{i-1,j} * b_{i,j}) * \bar{a}_{i,j},$$
$$g_{i,j} = (a_{i,j-1} * c_{i,j}) * \bar{a}_{i,j}.$$

Because $b_{i,j-1} * c_{i,j} \sim c_{i-1,j} * b_{i,j}$ it follows that the paths $f_{i,j-1} * g_{i,j}$ and $g_{i-1,j} * f_{i,j}$ are equivalent paths. Furthermore the equivalence is within U_1, U_2 or $U_1 \cap U_2$ according to whether $F(R_{i,j}) \subseteq U_1$, U_2 or $U_1 \cap U_2$ respectively. We therefore have

$$f_{i,j-1} \sim (g_{i-1,j} * f_{i,j}) * \bar{g}_{i,j},$$

i.e.

$$[f_{i,j-1}] = [g_{i-1,j}] \, [f_{i,j}] \, [\bar{g}_{i,j}].$$

Now express each of these elements as words in either S_1 or S_2, so that we get a relation

$$`[f_{i,j-1}]` = `[g_{i-1,j}]` \, `[f_{i,j}]` \, `[\bar{g}_{i,j}]`$$

within either $\pi(U_1, x_0)$ or $\pi(U_2, x_0)$ respectively. This relation must be a consequence therefore of the relations R_1 or R_2.

Suppose now that $1/k = t_{i(1)}$, then

$$[f_1] = [f_{1,0}] \, [f_{2,0}] \cdots [f_{i(1),0}]$$

and since f_1 is a closed path in $U_{\lambda(1)}$ based at x_0 we can use the relations $R_{\lambda(1)}$ to express $[f_{1,0}] \, [f_{2,0}], ..., [f_{i(1),0}]$ as words in $S_{\lambda(1)}$. Thus we get a relation

$$\alpha_1^{\epsilon(1)} = [f_1] = `[f_{1,0}]` \, `[f_{2,0}]` \cdots `[f_{i(1),0}]`$$

which is a consequence of the relations $R_{\lambda(1)}$. We can obtain similar relations for $\alpha_2^{\epsilon(2)}, \alpha_3^{\epsilon(3)}, ..., \alpha_k^{\epsilon(k)}$. Thus

$$\alpha = \alpha_1^{\epsilon(1)} \, \alpha_2^{\epsilon(2)} \cdots \alpha_k^{\epsilon(k)} = `[f_{1,0}]` \, `[f_{2,0}]` \cdots `[f_{m,0}]`$$

is a relation that is a consequence of R_1 and R_2. We rewrite this as

$$\alpha = (`[g_{0,1}]` \, `[f_{1,1}]` \, `[\bar{g}_{1,1}]`) (`[g_{1,1}]` \, `[f_{2,1}]` \, `[\bar{g}_{2,1}]`)$$
$$\cdots (`[g_{m-1,1}]` \, `[f_{m,1}]` \, `[\bar{g}_{m,1}]`)$$

giving a relation which is a consequence of R_1 and R_2. Rearranging brackets gives

$$\alpha = (`[g_{0,1}]`) \, `[f_{1,1}]` \, (`[\bar{g}_{1,1}]` \, `[g_{1,1}]`) \, `[f_{2,1}]` \, (`[\bar{g}_{2,1}]`$$
$$`[g_{2,1}]`) \cdots (`[\bar{g}_{m-1,1}]` \, `[g_{m-1,1}]`) \, `[f_{m,1}]` \, (`[\bar{g}_{m,1}]`).$$

Now $g_{0,1} = g_{m,1} = \epsilon_{x_0}$ so that $`[g_{0,1}]` = 1$ and $`[g_{m,1}]` = 1$ are trivial relations. The relation

$$`[\bar{g}_{j,1}]` \, `[g_{j,1}]` = 1$$

is also trivial if both $`[\bar{g}_{j,1}]`$ and $`[g_{j,1}]`$ are expressed as words in either

S_1 or S_2. However if $g_{j,1}$ is a path in $U_1 \cap U_2$ then it is possible that one of '$[\bar{g}_{j,1}]$', '$[g_{j,1}]$' is expressed as a word in S_1 and the other as a word in S_2. In this case the relation '$[\bar{g}_{j,1}]$' '$[g_{j,1}]$' = 1 is a consequence of the relations R_S. Thus we obtain the relation

$$\alpha = {}'[f_{1,1}]' \, '[f_{2,1}]' \, \cdots \, '[f_{m,1}]'$$

as a consequence of the relations R_1, R_2 and R_S.

By repetition of this process we arrive at the relation

$$\alpha = {}'[f_{1,n}]' \, '[f_{2,n}]' \, \cdots \, '[f_{m,n}]'$$
$$= {}'[\epsilon_{x_0}]' \, '[\epsilon_{x_0}]' \, \cdots \, '[\epsilon_{x_0}]'$$

as a consequence of the specified relations. Thus the original relation is a consequence of the specified relations. Thus Theorem 24.2, and hence also the Seifert–Van Kampen theorem, is proved.

Except for an example based upon the corollary that follows we shall leave calculations involving the Seifert–Van Kampen theorem until the next chapter.

24.3   Corollary

If $U_1 \cap U_2$ is simply connected then $\pi(X, x_0)$ is the group with generators $S_1 \cup S_2$ and relations $R_1 \cup R_2$.

The proof is obvious.

For an example using this corollary we shall look at a figure 8: thus X is the subspace of $\mathbb{R}^2$ consisting of $C_1 \cup C_2$ where

$$C_1 = \{ \, (x,y) \in \mathbb{R}^2 ; (x-1)^2 + y^2 = 1 \, \} \quad \text{and}$$
$$C_2 = \{ \, (x,y) \in \mathbb{R}^2 ; (x+1)^2 + y^2 = 1 \, \};$$

see Figure 24.2.

To apply the corollary we need to show that X is the union of open path connected subsets U_1, U_2 with $U_1 \cap U_2$ simply connected. Let, therefore, $U_1 = X - \{ \, x_1 \, \}$ and $U_2 = X - \{ \, x_2 \, \}$ where $x_1 = (-2,0)$, $x_2 = (2,0)$. Obviously U_1 and U_2 are both open and path connected; also $U_1 \cap U_2 =$

Figure 24.2

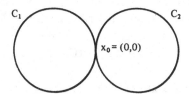

$X - \{ x_1, x_2 \}$ is path connected. Furthermore $U_1 \cap U_2$ is simply connected because it is homotopy equivalent to $\{ x_0 \}$; in fact $\{ x_0 \}$ is a strong deformation retract of X (see end of Chapter 13). Corollary 24.3 therefore applies. Now U_j and C_j are homotopy equivalent relative to x_0, for $j = 1, 2$ (in fact, again, C_j is a strong deformation retract of U_j). Thus $\pi(U_j, x_0)$ is a free group on one generator and so by Corollary 24.3 the fundamental group of a figure 8 is a free group on two generators.

It is not difficult to generalize this result to a collection of n circles joined at a single point, the result being that the fundamental group is a free group on n generators. We leave this as an (easy) exercise.

24.4　Exercises

(a) Let X_n be the union of n circles that intersect (pairwise and otherwise) at only one point x_0. Prove that $\pi(X_n, x_0)$ is the free group on n generators. (Induction?)

(b) Let X be the following subset of $\mathbb{R}^2$:

$$X = \{ (x,y) \in \mathbb{R}^2 ; -1 \le x, y \le 1 \text{ and } x \text{ or } y \in \mathbb{Z} \}.$$

Determine the fundamental group of X.

(c) Let Y be the complement of the following subset of $\mathbb{R}^2$:

$$\{ (x,0) \in \mathbb{R}^2 ; x \in \mathbb{Z} \}.$$

Prove that $\pi(Y, (1,1))$ is a free group on a countable set of generators.

(d) Let X be a Hausdorff space such that $X = A \cup B$, where A and B are each homeomorphic to a torus and $A \cap B = \{ x_0 \}$. Calculate $\pi(X, x_0)$. (Hint: For $x_0 \in A$ find a contractible neighbourhood C_A of x_0 in A then let $U_1 = B \cup C_A$. Similarly let $U_2 = A \cup C_B$ where C_B is a contractible neighbourhood of x_0 in B.)

(e) Let X be the space obtained from $S^{n-1} \times \mathbb{R}$ by removing k disjoint subsets each homeomorphic to the open n-disc D^n. What is the fundamental group of X?

(f) Let $X = \{ (x,y) \in \mathbb{R}P^n \times \mathbb{R}P^n ; x = x_0 \text{ or } y = x_0 \}$ where x_0 is some fixed point of $\mathbb{R}P^n$. (X is two copies of $\mathbb{R}P^n$ with one point x_0 in common.) Calculate $\pi(X, x_0)$. Is this group finite?

(g) Suppose that $X = U_1 \cup U_2$ with U_1, U_2 both open and path connected, and with $U_1 \cap U_2$ non-empty and path connected. Let $\varphi_1 : U_1 \cap U_2 \to U_1$ and $\psi_1 : U_1 \to X$ denote the inclusion maps. Prove that if U_2 is simply connected then $\psi_{1*} : \pi(U_1, x_0) \to \pi(X, x_0)$ is an epimorphism.

Furthermore prove that the kernel of ψ_{1*} is the smallest normal subgroup of $\pi(U_1, x_0)$ containing the image of $\varphi_{1*} \pi(U_1 \cap U_2, x_0)$.

(h) Suppose that $X = U_1 \cup U_2$ with U_1, U_2 and $U_1 \cap U_2$ open non-empty path connected subspaces. Prove that if U_2 and $U_1 \cap U_2$ are simply connected then the fundamental groups of U_1 and X are isomorphic.

(i) Let K be a compact subset of $\mathbb{R}^n$ with $\mathbb{R}^n - K$ path connected. Let $h: \mathbb{R}^n \to S^n - \{(1,0,0,...,0)\}$ be the homeomorphism given by

$$h(x_1, x_2, ..., x_n) = (\|x\|^2 - 1, 2x_1, 2x_2, ..., 2x_n) \left(\frac{1}{1 + \|x\|^2} \right)$$

Prove that if $x_0 \in \mathbb{R}^n - K$ then $\pi(\mathbb{R}^n - K, x_0)$ is isomorphic to $\pi(S^n - h(K), h(x_0))$. (Hint: $K \subseteq B_k(0)$ for some k. Consider $U_1 = h(\mathbb{R}^n - K)$, $U_2 = h(\mathbb{R}^n - B_k(0)) \cup \{(1, 0, ... 0)\}$ and then use (h) above.)

(j) Let X be a torus with one point removed. Show that the fundamental group of X is a free group on two generators.

(k) Show that the fundamental group of $\mathbb{R}P^2 - \{y\}$, where $y \in \mathbb{R}P^2$, is isomorphic to $\mathbb{Z}$.

(l) Let Y_n be the following subspace of $\mathbb{C}$:

$$Y_n = \{z \in \mathbb{C}; |z - j + \tfrac{1}{2}| = \tfrac{1}{2}, j = 1, 2, ..., n\}$$

where n is some positive integer. Calculate $\pi(Y_n, 0)$.

25

The Seifert-Van Kampen theorem: III Calculations

In this chapter we use the Seifert-Van Kampen theorem to calculate the fundamental groups of several spaces. For the first three examples the answer has been calculated before (as an exercise or corollary) using different methods.

We start by showing that the fundamental group of the torus T is isomorphic to $\mathbb{Z} \times \mathbb{Z}$. Represent T as a square region with edges identified as in Figure 25.1(b). We denote the edges that are identified by a_1 and a_2.

Let y be some point in the interior of the square region as indicated in Figure 25.1(c). Let $U_1 = T - \{ y \}$ and let $U_2 = T - (a_1 \cup a_2)$, i.e. U_2 is the interior of the square region. Obviously U_1 and U_2 are both open and path connected as is $U_1 \cap U_2$. Thus we can apply the Seifert-Van Kampen theorem. Let x_0, x_1 be the points indicated in Figure 25.1(c). (Note that x_1 appears four times in the diagram since these four points are identified to one point in T.) Finally let c be a circle, centre y, passing through x_0, and let d be the straight-line segment from x_0 to x_1 as indicated in Figure 25.1(c).

The 'edge' of the square region (Figure 25.2(a)) after identification gives a figure 8 in T (Figure 25.2(b); see also (a)). It is clearly a strong deformation retract of U_1.

If α_1 and α_2 denote closed paths in U_1, based at x_1, that go once along a_1 and a_2 respectively in the directions indicated then $\pi(U_1, x_1)$ is the free

Figure 25.1

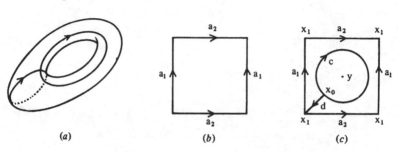

(a) (b) (c)

Figure 25.2

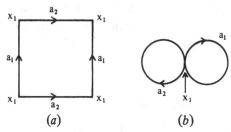

(a)   (b)

group on the generators $[\alpha_1]$, $[\alpha_2]$. Note that $[\alpha_1]$ and $[\alpha_2]$ are uniquely defined. Let δ denote a path in U_1 from x_0 to x_1 that corresponds to d, (i.e. $\delta: I \to d$ is a homeomorphism) then $\pi(U_1, x_0)$ is a free group on the generators $[\delta * \alpha_1 * \bar{\delta}]$, $[\delta * \alpha_2 * \bar{\delta}]$ which we abbreviate to A_1, A_2 respectively.

The single point space $\{ x_0 \}$ is a strong deformation retract of U_2 and so $\pi(U_2, x_0) = 1$. For the space $U_1 \cap U_2$ we see that the circle c is a strong deformation retract of $U_1 \cap U_2$. Thus if γ denotes a closed path in $U_1 \cap U_2$, based at x_0, which goes once around c in the direction indicated in Figure 25.1(b) then $\pi(U_1 \cap U_2, x_0)$ is a free group generated by $[\gamma]$.

The Seifert-Van Kampen theorem tells us that $\pi(T, x_0)$ is generated by $\{ A_1, A_2 \}$ and is subject to the following relation

$$'\varphi_{1*}[\gamma]' = '\varphi_{2*}[\gamma]'.$$

Now in U_1 we have

$$[\varphi_1 \gamma] = [\delta * \alpha_1 * \alpha_2 * \bar{\alpha}_1 * \bar{\alpha}_2 * \bar{\delta}]$$
$$= [\delta * \alpha_1 * \bar{\delta}][\delta * \alpha_2 * \bar{\delta}][\delta * \bar{\alpha}_1 * \bar{\delta}][\delta * \bar{\alpha}_2 * \bar{\delta}]$$

so that $'\varphi_{1*}[\gamma]' = A_1 A_2 A_1^{-1} A_2^{-1}$. On the other hand $'\varphi_{2*}[\gamma]' = 1$ so that $\pi(T, x_0)$ is the group with presentation $(\{ A_1, A_2 \}; \{ A_1 A_2 A_1^{-1} A_2^{-1} \})$ and hence $\pi(T, x_0)$ is isomorphic to $\mathbb{Z} \times \mathbb{Z}$.

For the next example consider the Klein bottle K. In many ways the calculation of the fundamental group of the Klein bottle is similar to that of the torus. Represent the Klein bottle K as in Figure 25.3(a) and use the notation of Figure 25.3(b).

Figure 25.3

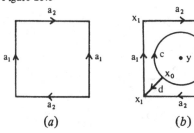

(a)   (b)

Let $U_1 = K - \{y\}$ and $U_2 = K - (a_1 \cup a_2)$; then $U_1, U_2, U_1 \cap U_2$ satisfy the conditions required in the Seifert–Van Kampen theorem. The 'edge' of the square region, after identification, is a figure 8 (see Figure 25.4) and this figure 8 is a strong deformation retract of U_1.

Figure 25.4

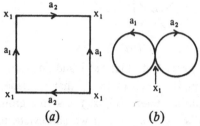

(a)   (b)

It follows that $\pi(U_1, x_0)$ is a free group generated by $\{[\alpha_1], [\alpha_2]\}$ where α_1, α_2 denote paths that correspond to a_1, a_2 respectively. If δ denotes the path corresponding to d then $\pi(U_1, x_0)$ is the free group generated by $[\delta * \alpha_1 * \bar{\delta}]$ and $[\delta * \alpha_2 * \bar{\delta}]$ which we abbreviate to A_1 and A_2 respectively.

The space U_2 is contractible and so $\pi(U_2, x_0) = 1$. Finally the circle c is a strong deformation retract of $U_1 \cap U_2$ so that $\pi(U_1 \cap U_2, x_0)$ is a free group generated by $[\gamma]$ where γ is a path in $U_1 \cap U_2$ that corresponds to c, i.e. goes once around c in the direction indicated.

In U_1 we have

$$[\varphi_1 \gamma] = [\delta * \alpha_1 * \alpha_2 * \bar{\alpha}_1 * \alpha_2 * \bar{\delta}]$$
$$= [\delta * \alpha_1 * \bar{\delta}][\delta * \alpha_2 * \bar{\delta}][\delta * \bar{\alpha}_1 * \bar{\delta}][\delta * \alpha_2 * \bar{\delta}]$$

so that '$\varphi_{1*}[\gamma]$' = $A_1 A_2 A_1^{-1} A_2$. On the other hand '$\varphi_{2*}[\gamma]$' = 1. Thus from the Seifert–Van Kampen theorem we immediately have that $\pi(K, x_0)$ is isomorphic to the group

$$\langle \{A_1, A_2\} ; \{A_1 A_2 A^{-1} A_2\} \rangle.$$

(See also Exercise 19.5(e).)

Figure 25.5

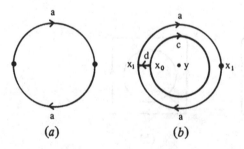

(a)   (b)

As a third example we show (again) that the fundamental group of the real projective plane $\mathbb{R}P^2$ is isomorphic to $\mathbb{Z}_2$. Represent $\mathbb{R}P^2$ as the identification space in Figure 25.5(a). Let x_0, x_1, y, c and d denote the points, circle and line as indicated in Figure 25.5(b).

Let $U_1 = \mathbb{R}P^2 - \{y\}$ and $U_2 = \mathbb{R}P^2 - a$; then $U_1, U_2, U_1 \cap U_2$ satisfy the conditions necessary in the Seifert-Van Kampen theorem. The curve a represents a circle in $\mathbb{R}P^2$ and it is a strong deformation retract of U_1. Thus $\pi(U_1, x_1)$ is a free group generated by $[\alpha]$ where α is a path in U_1 corresponding to a. If δ denotes the path from x_0 to x_1 corresponding to d then $\pi(U_1, x_0)$ is the free group generated by $[\delta * \alpha * \bar{\delta}] = A$ say.

The subspace U_2 is contractible to the point x_0 and so $\pi(U_2, x_0) = 1$. The circle c is a strong deformation retract of $U_1 \cap U_2$ so that $\pi(U_1 \cap U_2, x_0)$ is the free group with generator $[\gamma]$ where γ denotes a path in $U_1 \cap U_2$, based at x_0, that corresponds to c, i.e. goes once around c in the direction indicated. From the Seifert-Van Kampen theorem we deduce that $\pi(\mathbb{R}P^2, x_0)$ is the group with generator A and relation

$$'\varphi_{1*}[\gamma]' = '\varphi_{2*}[\gamma]'.$$

In U_1 we have

$$[\varphi_1 \gamma] = [\delta * \alpha * \alpha * \bar{\delta}] = [\delta * \alpha * \bar{\delta}][\delta * \alpha * \bar{\delta}],$$

so that $'\varphi_{1*}[\gamma]' = A^2$. Meanwhile $'\varphi_{2*}[\gamma]' = 1$ so that $\pi(\mathbb{R}P^2, x_0)$ is isomorphic to the group $\langle\{A\}; \{A^2\}\rangle$, i.e. to $\mathbb{Z}_2$.

For our next example let X be the space that consists of an n-sided polygonal region with all of its edges identified to one edge as indicated in Figure 25.6(a). Note that if $n = 2$ then X is $\mathbb{R}P^2$.

Using the notation in Figure 25.6(b) let $U_1 = X - \{y\}$ and $U_2 = X - a$. The spaces $U_1, U_2, U_1 \cap U_2$ are non-empty and path connected. The edges of the polygonal region form a circle a in X. It is a strong deformation retract of U_1 and so $\pi(U_1, x_1)$ is a free group generated by $[\alpha]$, where α is a closed

Figure 25.6

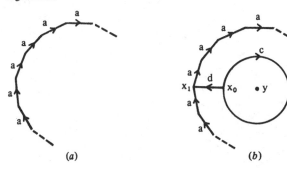

(a)   (b)

path, based at x_1, that corresponds to a. If δ is the path from x_0 to x_1 corresponding to d then $\pi(U_1,x_0)$ is a free group generated by $[\delta * \alpha * \bar{\delta}] = A$. The subspace U_2 is contractible so that $\pi(U_2,x_0) = 1$. Finally the circle c is a strong deformation retract of $U_1 \cap U_2$ so that $\pi(U_1 \cap U_2,x_0)$ is a free group generated by $[\gamma]$ where γ is a closed path in $U_1 \cap U_2$, based at x_0, that corresponds to c.

Applying the Seifert–Van Kampen theorem we see that $\pi(X,x_0)$ has one generator A and one relation

$$`\varphi_{1*}[\gamma]' = `\varphi_{2*}[\gamma]'.$$

The following is easy to see

$$[\varphi_1\gamma] = [\delta * \underbrace{\alpha * \,....\, * \alpha}_{n} * \bar{\delta}]$$

$$= [\delta * \alpha * \bar{\delta}]^n,$$

so that $`\varphi_{1*}[\gamma]' = A^n$. Meanwhile $`\varphi_{2*}[\gamma]' = 1$, so that $\pi(X,x_0)$ has presentation ({ A } ; { A^n }), i.e. $\pi(X,x_0)$ is isomorphic to the cyclic group $\mathbb{Z}_n$.

In all the preceding examples in this chapter the subspace U_2 is contractible. The next examples do not have this property. We shall look at three spaces at the same time. Let X_1,X_2,X_3 be the identification spaces illustrated in Figure 25.7. Notice that in X_3 the edge a_3 is not identified to any other edge. The notation that we shall use is illustrated in Figure 25.8.

Figure 25.7

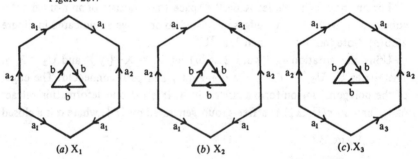

(a) X_1 (b) X_2 (c) X_3

Figure 25.8

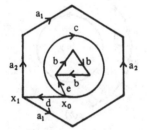

Let $U_{i,1} = X_i - b$ for $i = 1,2,3$. Let $U_{i,2} = X_i - (a_1 \cup a_2)$ for $i = 1,2$ and let $U_{3,2} = X_3 - (a_1 \cup a_2 \cup a_3)$. Then $U_{i,1}$, $U_{i,2}$, $U_{i,1} \cap U_{i,2}$ are all open and path connected subsets of X_i, $i = 1,2,3$. Figure 25.9 denotes the 'outer edges' of X_i after identification.

Figure 25.9

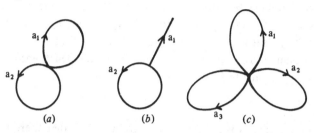

      (a)   (b)   (c)

In each case they are strong deformation retracts of $U_{i,1}$. It is therefore not difficult to see that $\pi(U_{i,1}, x_0)$ is a free group with generators:

$$\{ A_1 = [\delta * \alpha_1 * \bar\delta], A_2 = [\delta * \alpha_2 * \bar\delta] \} \qquad \text{if } i = 1,$$
$$\{ A_2 = [\delta * \alpha_2 * \bar\delta] \} \qquad \text{if } i = 2,$$
$$\{ A_1 = [\delta * \alpha_1 * \bar\delta], A_2 = [\delta * \alpha_2 * \bar\delta], A_3 = [\delta * \alpha_3 * \bar\delta] \} \text{ if } i = 3;$$

where in each case we use the obvious notation regarding $\alpha_1, \alpha_2, \alpha_3$ and δ.

The space $U_{i,2}$ contains a circle b which is a strong deformation retract of $U_{i,2}$, $i = 1,2,3$. Thus $\pi(U_{i,2}, x_0)$ is a free group with one generator $B = [\epsilon * \beta * \bar\epsilon]$ where β and ϵ are paths that correspond to b and e respectively.

The circle c in $U_{i,1} \cap U_{i,2}$ is a strong deformation retract of $U_{i,1} \cap U_{i,2}$ so that $\pi(U_{i,1} \cap U_{i,2})$ is a free group with one generator $[\gamma]$.

In $U_{i,1}$ we have the following:

For $i = 1$,
$$[\varphi_1 \gamma] = [\delta * \alpha_2 * \alpha_1 * \alpha_1 * \bar\alpha_2 * \alpha_1 * \bar\alpha_1 * \bar\delta]$$
$$= A_2 A_1{}^2 A_2{}^{-1}.$$

For $i = 2$,
$$[\varphi_1 \gamma] = [\delta * \alpha_2 * \alpha_1 * \bar\alpha_1 * \bar\alpha_2 * \alpha_1 * \bar\alpha_1 * \bar\delta]$$
$$= 1.$$

For $i = 3$,
$$[\varphi_1 \gamma] = [\delta * \alpha_2 * \alpha_1 * \alpha_1 * \bar\alpha_2 * \bar\alpha_3 * \bar\alpha_1 * \bar\delta]$$
$$= A_2 A_1{}^2 A_2{}^{-1} A_3{}^{-1} A_1{}^{-1}.$$

Within $U_{i,2}$ we have
$$[\varphi_2 \gamma] = [\epsilon * \beta * \beta * \beta * \bar\epsilon]$$
$$= B^3.$$

Using the Seifert-Van Kampen theorem we have the following results for the fundamental groups of X_i.

$$\pi(X_1,x_0) = \langle \{ A_1, A_2, B \}; \{ A_2 A_1{}^2 A_2^{-1} = B^3 \} \rangle,$$
$$\pi(X_2,x_0) = \langle \{ A_2, B \}; \{ B^3 = 1 \} \rangle,$$
$$\pi(X_3,x_0) = \langle \{ A_1, A_2, B \}; \emptyset \rangle.$$

The last result follows because $A_2 A_1{}^2 A_2^{-1} A_1^{-1} = B^3$ if and only if $A_3 = A_1^{-1} B^{-3} A_2 A_1{}^2 A_2^{-1}$, so that the group $\langle \{ A_1, A_2, A_3, B \}; \{ A_2 A_1{}^2 A_2^{-1} A_3^{-1} A_1^{-1} = B^3 \} \rangle$ is isomorphic to the group $\langle \{ A_1, A_2, B \}; \emptyset \rangle$.

Further calculations involving the Seifert-Van Kampen theorem will be given in subsequent chapters.

25.1   Exercises

(a) Suppose that G is a finite abelian group. Show that there is a space X_G whose fundamental group is isomorphic to G. (See also Exercise 19.5(b).)

(b) A space X is obtained from a pentagonal region by identifying its edges as indicated in Figure 25.10. Calculate the fundamental group of X.

Figure 25.10

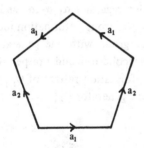

(c) Prove that if a subset W of $\mathbb{R}^3$ is homeomorphic to the open disc $\overset{\circ}{D}{}^2$ then W is not an open neighbourhood in $\mathbb{R}^3$ of any of its points. (Hint: If W is an open neighbourhood in $\mathbb{R}^3$ of $w \in W$ then there is a subset $U_1 \subseteq W$ with $w \in U_1$ and $U_1 \cong \overset{\circ}{D}{}^3$ by definition.)

(d) Let X be the double torus, i.e. the subspace of $\mathbb{R}^3$ depicted in Figure 11.7(e). Calculate the fundamental group of X.

(e) D_1, D_2 are two 2-discs with boundary circles S_1, S_2 respectively. The space X is the union of D_1 and D_2 with points in S_1 identified to points in S_2 by the rule

$\exp(2\pi it) \in S_1$ is identified with $\exp(2\pi int)$ in S_2

where n is some fixed positive integer. Prove that X is simply connected.

(f) Calculate the fundamental group of each of the identification spaces illustrated in Figure 25.11.

Figure 25.11

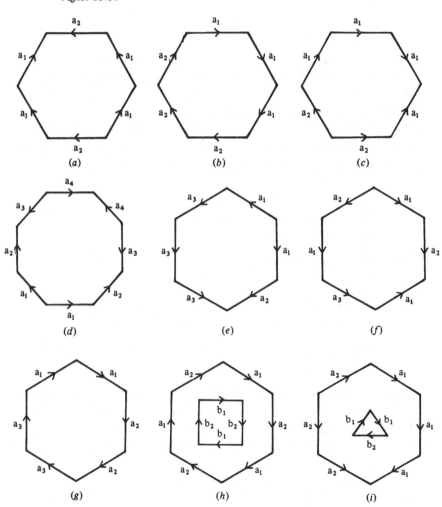

26

The fundamental group of a surface

It is now quite straight forward to calculate the fundamental group of a surface. Recall from Chapter 11 that any surface may be obtained from the sphere, torus and projective plane by taking connected sums. Recall also that the fundamental group of a torus is the group with two generators, say c_1, d_1, and one relation $c_1 d_1 c_1^{-1} d_1^{-1} = 1$, whereas the fundamental group of the projective plane is the group with one generator f_1 and one relation $f_1^2 = 1$. The general result we shall prove is:

26.1   Theorem
The fundamental group of the surface S,
$$S = S^2 \# mT \# n \mathbb{R} P^2$$
is the group with generators
$$c_1, d_1, c_2, d_2, ..., c_m, d_m, f_1, f_2, ..., f_n$$
and one relation
$$c_1 d_1 c_1^{-1} d_1^{-1} c_2 d_2 c_2^{-1} d_2^{-1} ... c_m d_m c_m^{-1} d_m^{-1} f_1^2 f_2^2 ... f_n^2 = 1.$$

Proof We may rewrite S as
$$S = X \cup H_1 \cup H_2 \cup ... \cup H_m \cup M_1 \cup M_2 \cup ... \cup M_n$$
where X is the sphere with $m + n = q$ disjoint open discs removed, $H_1, H_2, ...,$ H_m are handles (i.e. a torus with an open disc removed) and the $M_1, M_2, ...,$ M_n are Möbius strips (i.e. real projective planes with an open disc removed). If $b_1, b_2, ..., b_q$ denote the q boundary circles in X then we also have
$$X \cap H_i = b_i, \qquad i = 1, 2, ..., m,$$
$$X \cap M_j = b_{m+j}, \qquad j = 1, 2, ..., n.$$

Note that X is homeomorphic to the disc D^2 with $q - 1$ open discs removed. Let x_0 be the point in the interior of X as indicated in Figure 26.1. Also, let $x_1, x_2, ..., x_q$ be points in $b_1, b_2, ... b_q$ as indicated. Finally, let $a_1, a_2, ..., a_q$ be the curves between x_0 and $x_1, x_2, ..., x_q$ respectively as indicated in Figure 26.1.

Figure 26.1

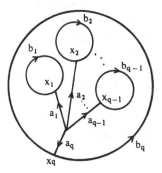

The subspace of X consisting of $a_1, a_2, ..., a_{q-1}$ and $b_1, b_2, ..., b_{q-1}$ is a strong deformation retract of X. It is then not difficult to see that the fundamental group is a free group on $q - 1$ generators. (In fact by shrinking $a_1, a_2, ..., a_{q-1}$ down to the point x_0 we see that X is homotopy equivalent, relative to $\{ x_0 \}$, to a union of $q - 1$ circles with exactly one point in common.) If we let $\alpha_1, \alpha_2, ..., \alpha_q$ be paths in X from x_0 to x_i corresponding to $a_1, a_2, ..., a_q$, and if we let $\beta_1, \beta_2, ..., \beta_q$ be closed paths, based at $x_1, x_2, ..., x_q$, corresponding to $b_1, b_2, ..., b_q$ then $\pi(X, x_0)$ is the free group generated by

$$B_1 = [\alpha_1 * \beta_1 * \bar{\alpha}_1], B_2 = [\alpha_2 * \beta_2 * \bar{\alpha}_2], ...,$$
$$B_{q-1} = [\alpha_{q-1} * \beta_{q-1} * \bar{\alpha}_{q-1}].$$

If B_q denotes $[\alpha_q * \beta_q * \bar{\alpha}_q]$ then

$$B_q^{-1} = B_1 B_2 ... B_{q-1},$$

i.e. $$B_1 B_2 ... B_{q-1} B_q = 1.$$

Thus, equally, $\pi(X, x_0)$ is the group with generators $B_1, B_2, ..., B_q$ and one relation $B_1 B_2 ... B_q = 1$. This formulation will be useful.

If we want a different base point, say x_i $(i = 1, 2, ..., q)$, then $\pi(X, x_i)$ is the group with generators $h_i(B_1), h_i(B_2), ..., h_i(B_q)$ and one relation $h_i(B_1 B_2 ... B_q)$ $= 1$ where $h_i: \pi(X, x_0) \to \pi(X, x_i)$ is the isomorphism given by $h_i([\theta]) = [\bar{\alpha}_i * \theta * \alpha_i]$. Note that $h_i(B_i) = [\beta_i]$.

Figure 26.2

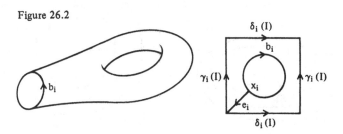

Now look at a handle H_i. From previous calculations we know that the fundamental group of H_i is a free group on two generators. With the notation of Figure 26.2, $\pi(H_i, x_i)$ is the free group generated by $C_i = [\epsilon_i * \gamma_i * \bar{\epsilon}_i]$ and $D_i = [\epsilon_i * \delta_i * \bar{\epsilon}_i]$ where ϵ_i is the path corresponding to the curve e_i in H_i and γ_i, δ_i are closed paths in H_i as indicated.

Note that the closed path β_i corresponding to b_i can be expressed in terms of C_i and D_i as

$$[\beta_i] = C_i D_i C_i^{-1} D_i^{-1}.$$

Consider the Möbius strip M_j. The fundamental group of M_j is a free group on one generator

$$F_j = [\epsilon_{j+m} * \varphi_j * \bar{\epsilon}_{j+m}]$$

where ϵ_{j+m} is the path corresponding to e_{j+m} in Figure 26.3 while φ_j is the closed path indicated there. Note also that $[\beta_{j+m}] = F_j^2$.

We shall combine the above results inductively in order to calculate the fundamental group of S. Define subspaces $X_0, X_1, ..., X_q$ of S as follows:

$$X_0 = X,$$
$$X_i = X_{i-1} \cup H_i \text{ for } i = 1, 2, ..., m,$$
$$X_{m+j} = X_{m+j-1} \cup M_j \text{ for } j = 1, 2, ..., n.$$

We shall show that the fundamental groups of these spaces X_i, X_{m+j} are as follows:

$\pi(X_i, x_0)$, $i = 0, 1, ..., m$, is the group with generators
$$c_1, d_1, c_2, d_2, ..., c_i, d_i, B_{i+1}, B_{i+2}, ..., B_q$$

and one relation

$$c_1 d_1 c_1^{-1} d_1^{-1} c_2 d_2 c_2^{-1} d_2^{-1} ... c_i d_i c_i^{-1} d_i^{-1} B_{i+1} B_{i+2} ... B_q = 1.$$

$\pi(X_{m+j}, x_0)$, $j = 0, 1, ..., n$, is the group with generators
$$c_1, d_1, c_2, d_2, ..., c_m, d_m, f_1, f_2, ..., f_j, B_{m+j+1}, B_{m+j+2}, ..., B_q$$

Figure 26.3

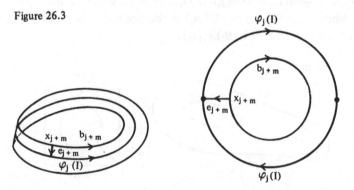

and one relation

$$c_1 d_1 c_1^{-1} d_1^{-1} c_2 d_2 c_2^{-1} d_2^{-1} \dots c_m d_m c_m^{-1} d_m^{-1} f_1^2 f_2^2 \dots f_j^2 B_{m+j+1} B_{m+j+1} \dots B_q = 1.$$

To prove the result we use the Seifert-Van Kampen theorem and this requires that we express X_k as a union of two open subsets. Although $X_k = X_{k-1} \cup Y_k$, where $Y_k = H_i$ or M_j for some i or j, unfortunately neither of these subspaces are open. However, recall from Chapter 11 that (because we have taken connected sums) there is an open neighbourhood N_k of b_k in S which is homeomorphic to $S^1 \times (-1,1)$, such that if $g_k : N_k \to S^1 \times (-1,1)$ is the homeomorphism then $g_k(b_k) = S^1 \times \{0\}$ and

$$g_k^{-1}(S^1 \times (-1,0]) \subseteq X \subseteq X_{k-1},$$
$$g_k^{-1}(S^1 \times [0,1)) \subseteq Y_k.$$

We therefore define

$$U_k = X_{k-1} \cup N_k \subseteq X_{k-1} \cup Y_k = X_k,$$

and

$$V_k = N_k \cup Y_k \subseteq X_{k-1} \cup Y_k \subseteq X_k.$$

Now U_k and V_k are open path connected subsets of X_k. Furthermore $U_k \cap V_k = N_k$ is path connected. We may now apply the Seifert-Van Kampen theorem to $X_k = U_k \cup V_k$, and $x_k \in b_k$ as base point. Of course, X_{k-1}, Y_k, b_k are strong deformation retracts of $U_k, V_k, U_k \cap V_k$ respectively so that

$$\pi(U_k, x_k) = \pi(X_{k-1}, x_k),$$
$$\pi(V_k, x_k) = \pi(Y_k, x_k),$$
$$\pi(U_k \cap V_k, x_k) = \pi(b_k, x_k) = \langle \{ [\beta_k] \} ; \emptyset \rangle.$$

It is quite easy to calculate the fundamental groups $\pi(X_k, x_0)$ by induction; this is illustrated by calculating $\pi(X_1, x_0)$. We have the following:

$$\pi(U_1, x_1) = \pi(X, x_1) = \langle \{ h_1(B_1), h_1(B_2), \dots, h_1(B_q) \} ;$$
$$\{ h_1(B_1 B_2 \dots B_q) = 1 \} \rangle,$$
$$\pi(V_1, x_1) = \pi(H_1, x_1) = \langle \{ C_1, D_1 \} ; \emptyset \rangle,$$
$$\pi(U_1 \cap V_1, x_1) = \pi(b_1, x_1) = \langle \{ [\beta_1] \} ; \emptyset \rangle.$$

By the Seifert-Van Kampen theorem $\pi(X_1, x_1) = \pi(U_1 \cup V_1, x_1)$ is the group with generators

$$C_1, D_1, h_1(B_1), h_1(B_2), \dots, h_1(B_q)$$

and relations

$$h_1(B_1 B_2 \dots B_q) = 1,$$
$$h_1(B_1) = C_1 D_1 C_1^{-1} D_1^{-1},$$

since $h_1(B_1) = [\beta_1]$ in X and $[\beta_1] = C_1 D_1 C_1^{-1} D_1^{-1}$ in H_1. Eliminating the generator $h_1(B_1)$ shows that $\pi(X_1, x_1)$ is the group with generators

$$C_1, D_1, h_1(B_2), h_1(B_3), ..., h_1(B_q)$$

and one relation

$$C_1 D_1 C_1^{-1} D_1^{-1} h_1(B_2 B_3 ... B_q) = 1.$$

It follows immediately that $\pi(X_1, x_0)$ is the group with generators

$$c_1, d_1, B_2, B_3, ..., B_q$$

and one relation

$$c_1 d_1 c_1^{-1} d_1^{-1} B_2 B_3 ... B_q = 1$$

where $c_1 = h_1^{-1}(C_1) = [\alpha_1 * \epsilon_1 * \gamma_1 * \bar{\epsilon}_1 * \bar{\alpha}_1]$ and $d_1 = h_1^{-1}(D_1) = [\alpha_1 * \epsilon_1 * \delta_1 * \bar{\epsilon}_1 * \bar{\alpha}_1]$.

How to proceed is obvious and left for the reader.

It is not immediately obvious as to how distinct the groups listed in Theorem 26.1 are. We therefore abelianize them, where if $G = \langle S_1; R \rangle$ then G *abelianized* is

$$AG = \langle S_1; R \cup \{ xyx^{-1}y^{-1}; x, y \in S_1 \} \rangle,$$

i.e. we add the extra relations $xy = yx$ for all $x, y \in G$.

Suppose that $n = 0$, i.e. $S = S^2 \# mT$; then $A\pi(S, x_0)$ is the group with generators

$$S_m = \{ c_1, d_1, c_2, d_2, ..., c_m, d_m \}$$

and relations $\{ r_m = 1 \} \cup \{ xy = yx; x, y \in S_m \}$ where

$$r_m = c_1 d_1 c_1^{-1} d_1^{-1} c_2 d_2 c_2^{-1} d_2^{-1} ... c_m d_m c_m^{-1} d_m^{-1}.$$

In particular we have the relation $c_1 d_1 = d_1 c_1$, so that the relation $r_m = 1$ is a consequence of the relations $\{ xy = yx; x, y \in S_m \}$. Thus $A\pi(S, x_0)$ is the group

$$\langle S_m; \{ xy = yx; x, y, \in S_m \} \rangle$$

and it is not difficult to see that $A\pi(S, x_0) \cong \mathbb{Z}^{2m}$.

If $n \geq 1$, so that $S = S^2 \# mT \# n\mathbb{R}P^2$, then $A\pi(S, x_0)$ is the group with generators

$$S_{m+n} = \{ c_1, d_1, c_2, d_2, ..., c_m, d_m, f_1, f_2, ..., f_n \}$$

and relations $\{ r_{m+n} = 1 \} \cup \{ xy = yx; x, y \in S_{m+n} \}$ where

$$r_{m+n} = c_1 d_1 c_1^{-1} d_1^{-1} c_2 d_2 c_2^{-1} d_2^{-1} ... c_m d_m c_m^{-1} d_m^{-1} f_1^2 f_2^2 ... f_n^2.$$

The relation $r_{m+n} = 1$ is a consequence of the relations $\{ xy = yx; x, y \in S_{m+n} \}$ and $\{ (f_1 f_2 ... f_n)^2 = 1 \}$. Furthermore the relation $\{ (f_1 f_2 ... f_n)^2 = 1 \}$ is a consequence of the relations $\{ r_{m+n} = 1 \} \cup \{ xy = yx; x, y \in S_{m+n} \}$. Thus

$$A\pi(S,x_0) = \langle S_{m+n}; \{ \ xy = yx; x,y \in S_{m+n} \ \} \cup$$
$$\{ \ (f_1 f_2 ... f_n)^2 = 1 \ \} \rangle.$$

Now any element of $A\pi(S,x_0)$ may be written as

$$c_1^{a(1)} d_1^{b(1)} c_2^{a(2)} d_2^{b(2)} ... c_m^{a(m)} d_m^{b(m)} f_1^{e(1)} f_2^{e(2)} ... f_n^{e(n)}$$

where $a(i), b(i), e(i) \in \mathbb{Z}$. This may be rewritten as

$$c_1^{a(1)} d_1^{b(1)} c_2^{a(2)} d_2^{b(2)} ... c_m^{a(m)} d_m^{b(m)} f_1^{e(1)-e(n)} f_2^{e(2)-e(n)} ...$$
$$f_{n-1}^{e(n-1)-e(n)} (f_1 f_2 ... f_n)^{e(n)}$$

and then we can see that $A\pi(S,x_0) \cong \mathbb{Z}^{2m+n-1} \times \mathbb{Z}_2$.

26.2 Corollary
The abelianized fundamental group
(i) of an orientable surface of genus m ($m \geq 0$) is $\mathbb{Z}^{2m}$;
(ii) of a non-orientable surface of genus n ($n \geq 1$) is $\mathbb{Z}^{n-1} \times \mathbb{Z}_2$.

This corollary proves that no two of the surfaces listed in Theorem 11.3 are homeomorphic.

The next result may be viewed as the *basic result relating surfaces and fundamental groups.*

26.3 Corollary
Two surfaces are homeomorphic if and only if their (abelianized) fundamental groups are isomorphic.

This follows from the classification theorem of surfaces (Chapter 11) and Corollary 26.2.

As another corollary we have

26.4 Corollary
A surface is simply connected if and only if it is homeomorphic to the sphere S^2.

Corollary 26.2 could be used as a way of deciding whether or not a space is a surface:

26.5 Corollary
Let X be a space with $x_0 \in X$. If $A\pi(X,x_0)$ is not of the form $\mathbb{Z}^{2m}$ or $\mathbb{Z}^{n-1} \times \mathbb{Z}_2$ then X is not a surface.

In Chapter 11 we gave an alternative description of surfaces in terms of quotient spaces of polygonal regions. We leave it as an exercise for the reader to recalculate the fundamental group of a surface using this alternative description.

26.6   Exercises

(a) Let M be the quotient space of a 4m-sided polygonal region ($m \geq 1$) with identifications as indicated in Figure 26.4(a), i.e. M is an orientable surface of genus m. Prove directly (using the Seifert–Van Kampen theorem) that the fundamental group of M is the group with generators

$$A_1, B_1, A_2, B_2, ..., A_m, B_m$$

and one relation

$$A_1 B_1 A_1^{-1} B_1^{-1} A_2 B_2 A_2^{-1} B_2^{-1} ... A_m B_m A_m^{-1} B_m^{-1} = 1.$$

(b) Let M be the quotient space of a 2n-sided polygonal region ($n \geq 1$) with identifications as indicated in Figure 26.4(b), i.e. M is a non-orientable surface of genus n. Prove (using the Seifert–Van Kampen theorem) that the fundamental group of M is the group with generators

$$A_1, A_2, ..., A_n$$

and one relation

$$A_1^2 A_2^2 ... A_n^2 = 1.$$

(c) Prove that if M_1 and M_2 are connected n-manifolds with $n > 2$ then the fundamental group of $M_1 \# M_2$ is isomorphic to the group $\langle S_1 \cup S_2; R_1 \cup R_2 \rangle$ where the fundamental group of M_i is $\langle S_i, R_i \rangle$ for $i = 1, 2$.

(d) Prove that if G_n is a free group on n generators then there is a 4-manifold M_n with fundamental group G_n. (Hint: Find M_1.)

Figure 26.4

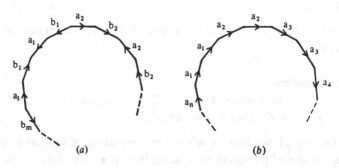

(a)   (b)

27

Knots: I Background and torus knots

A *knot* is a subspace of $\mathbb{R}^3$ that is homeomorphic to the circle S^1. Some examples are given in Figure 27.1. Although all the spaces in Figure 27.1 are homeomorphic to each other (since each is homeomorphic to a circle by definition), our intuition tells us that within $\mathbb{R}^3$ they are not the same. Thus, if we make models of knots using string, then within our three-dimensional world we could not, for example, create knot (*c*) from knot (*a*) in Figure 27.1, unless we cut the string at some stage. This is because knot (*c*) is 'knotted' while knot (*a*) is 'unknotted'. It is reasonable to say that a knot is unknotted if we can move it continuously within 3-space to the knot (*a*) of Figure 27.1. This suggests that, in addition to the knot being moved continuously, the surrounding 3-space is also moved continuously. We are thus led to the following definition. A knot K is *unknotted* if there is a homeomorphism h: $\mathbb{R}^3 \to \mathbb{R}^3$ such that h(K) is the standard circle $\{ (x,y,0) \in \mathbb{R}^3 ; x^2 + y^2 = 1 \}$ in $\mathbb{R}^2 \subset \mathbb{R}^3$. Thus knots (*a*) and (*b*) of Figure 27.1 are unknotted while the others are not; at least practical experience or intuition tells us so. Later on in this chapter we shall prove that the knots (*c*), (*d*) and (*g*) of Figure 27.1 are not unknotted. In the next chapter we shall be in a position to prove that all the remaining knots of Figure 27.1 are not unknotted.

Before continuing, the reader may have wondered why we have defined a knot to be a subspace of $\mathbb{R}^3$. (Is it only because we appear to live in a three-dimensional world?) Why not define a knot K to be a subspace of $\mathbb{R}^n$ such that K is homeomorphic to S^1. Obviously n has to be at least 2 (why? – Corollary 10.3!). However, if $n \neq 3$ then there is a homeomorphism h: $\mathbb{R}^n \to \mathbb{R}^n$ such that h(K) is the standard circle in $\mathbb{R}^n$. We shall not prove this result. For n = 2 this is the famous Schönflies theorem. For $n \geq 4$ the result tells us that if we lived in a four (or more)-dimensional world then we could unknot all knots. Intuitively this should be clear; the extra dimension gives us room to push one piece of the string 'through' another. This explains why we defined knots as homeomorphic images of S^1 in $\mathbb{R}^3$. We could also

Figure 27.1. Some knots.

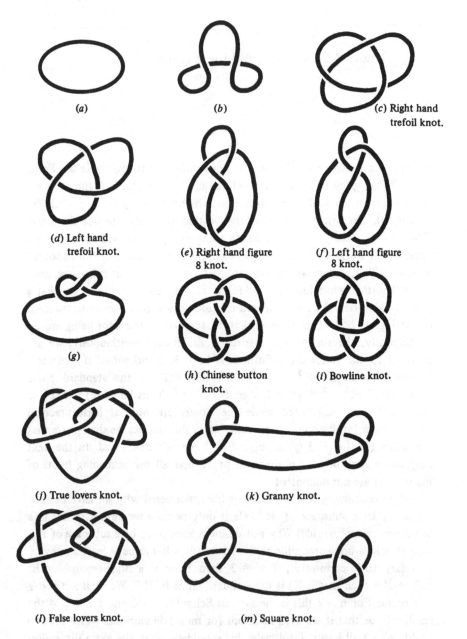

(a)

(b)

(c) Right hand
trefoil knot.

(d) Left hand
trefoil knot.

(e) Right hand figure
8 knot.

(f) Left hand figure
8 knot.

(g)

(h) Chinese button
knot.

(i) Bowline knot.

(j) True lovers knot.

(k) Granny knot.

(l) False lovers knot.

(m) Square knot.

consider subspaces of $\mathbb{R}^{n+1}$ which are homeomorphic to S^{n-1}. The question makes sense and leads to some interesting mathematics, but this is beyond the scope of this book.

Going back to knots (in $\mathbb{R}^3$) we have defined what it means for a knot to be unknotted. More generally we say that two knots K_1, K_2 are *similar* if there is a homeomorphism h: $\mathbb{R}^3 \to \mathbb{R}^3$ such that $h(K_1) = K_2$. For example in Figure 27.1, knots (a) and (b) are similar, knots (c), (d) and (g) are similar, knots (e) and (f) are similar, etc. That (a) and (b) are similar is obvious, also, that (g) and (c) are similar is easy to see. To see that (c) and (d) are similar, place one directly above the other. A mirror in between the two provides the required homeomorphism of $\mathbb{R}^3$. A similar homeomorphism works for the pair (e), (f). There is another way to see that knots (e) and (f) are similar. This is depicted by the sequences of diagrams in Figure 27.2. The reader is advised to make knot (e) of Figure 27.1 out of some string and produce the sequence of knots depicted in Figure 27.2.

Figure 27.2

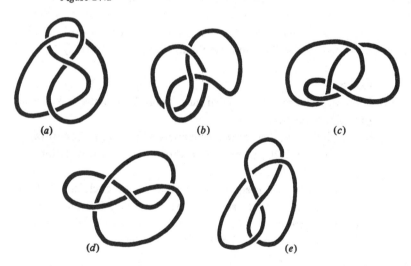

(a)   (b)   (c)

(d)   (e)

Note however that physical experimentation (try it) tells us that the left handed and right handed *trefoil* knots (Figure 27.1 (c), (d)) are not the same in the sense that we cannot move the left handed trefoil knot within 3-space to create the right handed trefoil knot. In fact we need a mirror to get from one to the other. A mirror takes a 'right hand frame' in $\mathbb{R}^3$ into a 'left hand frame' (Figure 27.3) and there is no way, within $\mathbb{R}^3$, of moving a right hand frame into a left hand frame.

Figure 27.3

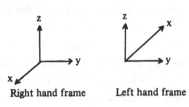

Right hand frame   Left hand frame

We say that a homeomorphism h: $\mathbb{R}^3 \to \mathbb{R}^3$ is *orientation preserving* if h sends a right hand frame to a right hand frame. Two knots K_1 and K_2 are *equivalent* if there is an orientation preserving homeomorphism h: $\mathbb{R}^3 \to \mathbb{R}^3$ such that $h(K_1) = K_2$. Thus (intuitively at least) the left hand and right hand trefoil knots are not equivalent. On the other hand the left hand and right hand figure 8 knots are equivalent, as Figure 27.2 shows. The notion of knots being equivalent agrees well with the physical notion of knots being the same. In fact one could prove that two knots K_1, K_2 are *equivalent* if and only if there is a homeomorphism h: $\mathbb{R}^3 \to \mathbb{R}^3$ and a real number $k > 0$ such that $h(K_1) = K_2$ and $h(x) = x$ whenever $\|x\| \geq k$. This result has strong physical connotations which we shall leave for the reader to work out. We shall neither prove nor use the result just mentioned (it is not trivial).

27.1   Exercises

(a) Show that the relations 'similar' and 'equivalent' between knots are equivalence relations.

(b) Let h: $\mathbb{R}^3 \to \mathbb{R}^3$ be a linear mapping (i.e. $h(\lambda a + \mu b) = \lambda h(a) + \mu h(b)$ for $\lambda, \mu \in \mathbb{R}$, a, b $\in \mathbb{R}^3$). Prove that h is orientation preserving if and only if det h = +1.

(c) Prove that a knot K is equivalent to the standard circle in $\mathbb{R}^3$ if and only if K is similar to the standard circle in $\mathbb{R}^3$. (Hint: The standard circle is 'symmetric'.)

(d) Find examples (if possible) of knots K such that
   (i) $K \subset S^2 \subset \mathbb{R}^3$.
   (ii) $K \subset$ torus $\subset \mathbb{R}^3$,
   (iii) $K \subset$ double torus $\subset \mathbb{R}^3$.

(e) Let K be a knot in $\mathbb{R}^3$ that consists of a finite number, say k, of straight-line segments. For what values of k $(1 \leq k \leq 10)$ can you find a knot K which is not unknotted?

(f) Let p: $\mathbb{R}^3 \to \mathbb{R}^2$ denote the natural projection of $\mathbb{R}^3$ onto $\mathbb{R}^2$ (i.e. $p(x_1, x_2, x_3) = (x_1, x_2) \in \mathbb{R}^2 \subset \mathbb{R}^3$). A *crossing point* of a knot K is a point x $\in \mathbb{R}^2$ such that $p^{-1}(x) \cap K$ consists of two or more points. It is a *double point* if $p^{-1}(x) \cap K$ consists of two points. In this case we say that it is not a proper double point if by moving the knot slightly the double point disappears; see Figure 27.4.

Consider knots such that all crossing points are proper double crossing points. Find all such knots with one, two, three, four, five or six crossing points.

If K_1 and K_2 are similar knots then there is a homeomorphism between $\mathbb{R}^3 - K_1$ and $\mathbb{R}^3 - K_2$. Hence the fundamental groups of the complements of two similar knots are isomorphic. Choose any point $x_0 \in \mathbb{R}^3 - K$; we call $\pi(\mathbb{R}^3 - K, x_0)$ *the group of the knot* K. Knots that are similar but which are not equivalent still have isomorphic groups. This explains why we look at similar knots and not equivalent knots.

27.2   Theorem
The group of an unknotted knot is isomorphic to $\mathbb{Z}$.

Proof Let $K = \{ (x,y,z) \in \mathbb{R}^3 ; x^2 + y^2 = 1, z = 0 \}$ be the standard circle in $\mathbb{R}^3$. Let $\epsilon > 0$ be an arbitrary small real number and define subspaces X, Y of $\mathbb{R}^3 - K$ by

$$X = \{ (x,y,z) \in \mathbb{R}^3 ; x < \epsilon \} \cap \mathbb{R}^3 - K,$$
$$Y = \{ (x,y,z) \in \mathbb{R}^3 ; x > -\epsilon \} \cap \mathbb{R}^3 - K.$$

It is easy to see that the subspace

$$\{ (x,y,z) \in \mathbb{R}^3 ; x = 0, y \geq 0 \} - \{ (0,1,0) \}$$

is a strong deformation retract of X and of Y. Thus if O is the origin of $\mathbb{R}^3$ then the fundamental groups $\pi(X,O)$, $\pi(Y,O)$ are both isomorphic to $\mathbb{Z}$ with generators $[\alpha_X]$, $[\alpha_Y]$, where $\alpha_X \colon I \to X$ and $\alpha_Y \colon I \to Y$ are paths each defined by

$$t \to (0, 1 - \cos(2\pi t), \sin(2\pi t)).$$

Figure 27.4

Proper double crossing point   Improper double crossing point

The subspace

$$\{ (x,y,z); x = 0 \} - \{ (0,1,0), (0,-1,0) \}$$

is a strong deformation retract of $X \cap Y$ so that $\pi(X \cap Y, O)$ is the free group on two generators $[\beta_1]$ and $[\beta_{-1}]$ defined by

$$\beta_1(t) = (0, 1 - \cos(2\pi t), \sin(2\pi t)),$$
$$\beta_{-1}(t) = (0, -1 + \cos(2\pi t), \sin(2\pi t)).$$

If $\varphi_X : X \cap Y \to X$, $\varphi_Y : X \cap Y \to Y$ denote the natural inclusions then it is clear that

$$\varphi_X \beta_1 = \alpha_X, \qquad \varphi_Y \beta_1 = \alpha_Y,$$
$$\varphi_X \beta_{-1} \sim \alpha_X, \qquad \varphi_Y \beta_{-1} \sim \alpha_Y.$$

All conditions on X, Y, $X \cap Y$ that are necessary for the application of the Seifert-Van Kampen theorem are satisfied and so by using this theorem we deduce that the fundamental group $\pi(\mathbb{R}^3 - K, O)$ is a free group on one generator $[\alpha]$, where $\alpha : I \to \mathbb{R}^3 - K$ is given by

$$\alpha(t) = (0, 1 - \cos(2\pi t), \sin(2\pi t)).$$

Our next goal is to show that not all knots are unknotted. We do this by calculating the group of the trefoil knots and related knots. The trefoil knots belong to a group of the so-called torus knots. This is a large class of knots which occur as simple closed curves on a torus in $\mathbb{R}^3$. We think of the torus as $S^1 \times S^1$ with a point in $S^1 \times S^1$ given by a pair $(\exp(i\varphi), \exp(i\theta))$ where $0 \leq \varphi, \theta < 2\pi$. It is convenient to think of $\mathbb{R}^3$ as $\mathbb{C} \times \mathbb{R}$ and to use polar coordinates $(r, \theta) \equiv re^{i\theta}$ in $\mathbb{C}$. Thus a point in $\mathbb{R}^3$ is represented by a triple (r, θ, z). In these terms we have a continuous map $f: S^1 \times S^1 \to \mathbb{R}^3$ given by

$$f(\exp(i\varphi), \exp(i\theta)) = (1 + \tfrac{1}{2}\cos\varphi, \theta, \tfrac{1}{2}\sin\varphi),$$

and $S^1 \times S^1$ is homeomorphic to the image of $f(S^1 \times S^1)$. See Figure 27.5 and Figure 5.4.

Figure 27.5

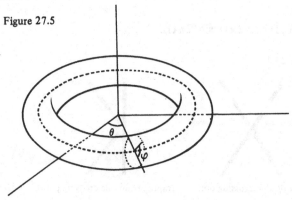

Figure 27.6

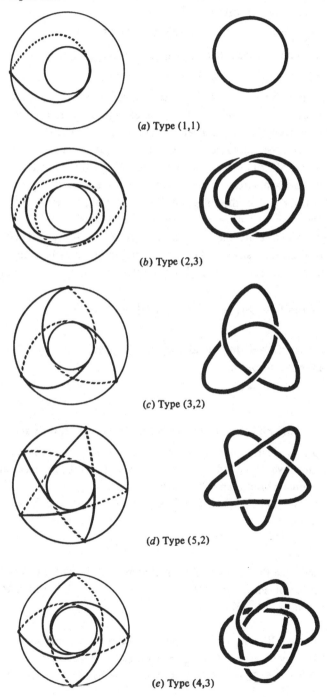

(*a*) Type (1,1)

(*b*) Type (2,3)

(*c*) Type (3,2)

(*d*) Type (5,2)

(*e*) Type (4,3)

Let m,n be a pair of coprime positive integers. Define $K_{m,n}$ to be the following subset of the torus in $\mathbb{R}^3$

$$K_{m,n} = \{ \ f(\exp(2\pi imt), \exp(2\pi int)); t \in I \ \}.$$

It is not difficult to check that the map g: $S^1 \to K_{m,n}$ defined by

$$g(\exp(2\pi it)) = f(\exp(2\pi imt), \exp(2\pi int))$$

is a homeomorphism, so that $K_{m,n}$ is a knot. We call $K_{m,n}$ the *torus knot of type* (m,n). See Figure 27.6 for some examples. There are two standard circles in the torus; these are given by $f(\exp(2\pi it), 1)$ and $f(1,\exp(2\pi it))$. A torus knot of type (m,n) goes n times around the torus in the direction of the circle $f(\exp(2\pi it), 1)$ and m times around the torus in the direction of the other standard circle. If we think of a torus T as the quotient space $\mathbb{R}^2/\mathbb{Z}^2$ then $K_{m,n}$ is the image in T of the straight line in $\mathbb{R}^2$ that goes through the origin at a slope n/m to the horizontal axis.

In order to calculate the group of a torus knot $K = K_{m,n}$ it will be convenient to thicken K slightly. Let a be a small positive real number which is smaller than $|\frac{1}{2} \sin(\pi/n)|$; for $0 \leq b \leq a$ define K_b to be the following subset of $\mathbb{R}^3$ (= $\mathbb{C} \times \mathbb{R}$):

$$\{ \ (x+1+\tfrac{1}{2} \cos(2\pi mt), 2\pi nt, y+\tfrac{1}{2} \sin(2\pi mt));$$
$$0 \leq t \leq 1, x^2 + y^2 \leq b^2 \ \}.$$

It is clear that $K_0 = K$. In general if $D_b = \{ \ (x,y) \in \mathbb{R}^2 ; x^2 + y^2 \leq b^2 \ \}$ then there is a homeomorphism

$$h \colon S^1 \times D_b \to K_b$$

given by

$$h(\exp(2\pi it), (x,y)) = (x+1+\tfrac{1}{2} \cos(2\pi mt), 2\pi nt, y+\tfrac{1}{2} \sin(2\pi mt)).$$

(The map h is clearly continuous and surjective. It is not difficult to check that h is injective, the condition $b \leq a < |\frac{1}{2} \sin(\pi/n)|$ being necessary. That h is a homeomorphism then follows from Theorem 8.8.)

It turns out that $\mathbb{R}^3 - K$ and $\mathbb{R}^3 - K_b$ are homeomorphic for each b such that $0 \leq b < a$.

27.3   Theorem
Suppose that a is a positive real number and that for $0 \leq b \leq a$ there is a homeomorphism h: $S^1 \times D_b \to K_b \subset \mathbb{R}^3$. If $0 \leq b < a$ then $\mathbb{R}^3 - K_b$ is homeomorphic to $\mathbb{R}^3 - K_0$.

Proof Define φ: $\mathbb{R}^3 - K_b \to \mathbb{R}^3 - K_0$ in the following way: If $x \notin \mathring{K}_a - K_b$ then let $\varphi(x) = x$, while if $x \in K_a - K_b$ then we can write $x = h(z, r\exp(i\theta))$ with $b < r \leq a$; we define $\varphi(x)$ in this case to be $h(z,(r-b)(a/(a-b))\exp(i\theta))$.

Similarly, define $\psi\colon \mathbb{R}^3-K_0 \to \mathbb{R}^3-K_b$ in the following way: If $x \notin \overset{\circ}{K}_a - K_0$ then $\psi(x) = x$, while if $x \in K_a-K_0$ then $x = h(z,r\exp(i\theta))$ with $0 < r \leq a$, and we define $\psi(x)$ to be $h(z,(r(a-b)/a + b)\exp(i\theta))$. We leave it for the reader to check that $\psi\varphi = 1$, $\varphi\psi = 1$ and that φ,ψ are continuous.

27.4   Theorem

The group of a torus knot K of type (m,n) is the group on two generators a,b and one relation $a^n = b^m$.

Proof Define $f\colon \mathbb{C} \times S^1 \to \mathbb{R}^3 \ (= \mathbb{C} \times \mathbb{R})$ by

$$(r \exp(i\varphi), \exp(i\theta)) \to (1 + \tfrac{1}{2}r \cos\varphi, \theta, \tfrac{1}{2}r \sin\varphi)$$

Clearly f is continuous. Also $f(S^1 \times S^1)$ gives us the torus in $\mathbb{R}^3$ that we described earlier on. Choose $\epsilon > 0$ so that $2\epsilon < a$ where a, as before, is smaller than $|\tfrac{1}{2} \sin(\pi/n)|$. Define subspaces X and Y by

$$X = \{\ f(r \exp(i\varphi), \exp(i\theta)); 0 \leq \varphi, \theta \leq 2\pi, 0 \leq r < 1 + \epsilon\ \} - K_{2\epsilon},$$

$$Y = \mathbb{R}^3 - \{\ f(r \exp(i\varphi), \exp(i\theta)); 0 \leq \varphi, \theta \leq 2\pi, 0 \leq r < 1 - \epsilon\ \}$$
$$- K_{2\epsilon}$$

Thus $X \cup Y = \mathbb{R}^3 - K_{2\epsilon} \cong \mathbb{R}^3-K$. Let $x_0 \in X \cap Y$.

It is easy to see that $Z = \{\ f(0,\exp(i\theta)); 0 \leq \theta \leq 2\pi\ \}$ is a strong deformation retract of X and that $\mathbb{R}^3-Z$ is a strong deformation retract of Y. See Figure 27.7 which shows the various regions involved (for $n = 3$) intersected with the half-plane H of $\mathbb{R}^3$ given by points (r,θ,z) with θ fixed (say $\theta = 0$). Of course Z is the standard circle in $\mathbb{R}^3$ and so $\pi(X,x_0)$ and $\pi(Y,x_0)$ are free groups on one generator $[\alpha]$, $[\beta]$ respectively. Here α represents a closed path in X that goes from x_0 to the circle Z along an arc a_X say, once around Z and then back along the arc a_X to x_0. The generator

Figure 27.7

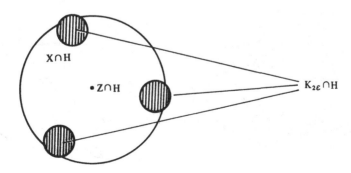

$[\beta]$ is represented by a closed path β in Y that goes from x_0 along an arc a_Y to the circle

$$\{ (1+\tfrac{3}{4}\cos(2\pi t), 0, \tfrac{3}{4}\sin(2\pi t)); t \in I \},$$

once around this circle and then back along the arc a_Y to x_0.

The space $X \cap Y$ is precisely

$$\{ f(r\exp(i\varphi), \exp(i\theta)); 0 \le \varphi, \theta \le 2\pi, 1-\epsilon < r < 1+\epsilon \} - K_{2\epsilon},$$

and contains the subspace

$$W = \{ f(\exp(2\pi i(mt+\delta)), \exp(2\pi int)); 0 \le t \le 1 \}$$

for a suitable δ (say $\delta = \tfrac{1}{2}n$) which is a strong deformation retract of $X \cap Y$. (Try drawing some examples.) This subspace W is a circle and so we deduce that $\pi(X \cap Y, x_0)$ is a free group on one generator $[\gamma]$, where γ is a closed path in $X \cap Y$ that goes once around W (if we choose $x_0 \in W$).

Let $\varphi_X : X \cap Y \to X$, $\varphi_Y : X \cap Y \to Y$ denote the natural inclusions; then it is not difficult to see that

$$\varphi_{X*}[\gamma] = [\alpha]^n \text{ or } [\alpha]^{-n},$$
$$\varphi_{Y*}[\gamma] = [\beta]^m \text{ or } [\beta]^{-m},$$

so that replacing α and/or β by $\bar{\alpha}$ and/or $\bar{\beta}$ respectively we have

$$\varphi_{X*}[\gamma] = [\alpha]^n, \varphi_{Y*}[\gamma] = [\beta]^m.$$

See Figure 27.8 which illustrates the case m = 3, n = 2.

The spaces $X, Y, X \cap Y$ satisfy the necessary conditions for the application of the Seifert–Van Kampen theorem from which the result follows immediately. Note that the Seifert–Van Kampen theorem could not have been applied if in the above we had actually had $\epsilon = 0$.

Having calculated the group of a torus knot it is not clear as to whether the group is trivial or not. For surfaces we had a similar problem which we solved by abelianizing the group. However, it turns out that the abelianized group of a knot group is always isomorphic to $\mathbb{Z}$ (see Exercise 27.7(d) for

Figure 27.8

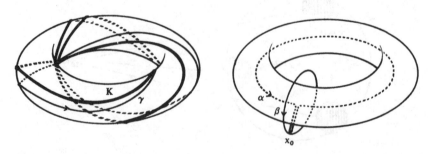

hints on how to prove this for torus knots; in general see Corollary 28.4). We must therefore look for some other way of deciding whether torus knots are knotted or not.

27.5   Lemma
The torus knot of type (3,2) is not unknotted.

Proof The group of this knot is $G = \langle \{a,b\}; \{a^2 = b^3\} \rangle$. Define G' by adding a few relations to G

$$G' = \langle \{a,b\}; \{a^2 = b^3, a^2 = 1, ab = b^{-1}a\} \rangle.$$

There is an obvious epimorphism $G \to G'$ which on the generators sends a to a and b to b.

It is easy to see that G' is isomorphic to the symmetric group on three letters, i.e.

$$G' = \{1,a,b,ab,b^2,ab^2 ; ab = b^{-1}a\}.$$

This group is non-abelian which means that G itself is non-abelian since the image under a homomorphism of an abelian group is abelian. Thus G is not isomorphic to Z, which proves that the torus knot of type (3,2) is not unknotted.

We could produce similar arguments for the other torus knots, but this would only show that they are not unknotted. We would like to know if any of them are similar to each other. We need another trick to show that they are different.

27.6   Theorem
If two torus knots of type (m,n), (m',n') with $m,n,m',n' > 1$ are similar then $\{m,n\} = \{m',n'\}$, i.e. $m = m'$ and $n = n'$ or $m = n'$ and $n = m'$. In particular if $m,n > 1$ then no torus knot of type (m,n) can be unknotted. Also, there are infinitely many different knots.

Proof The group theoretical argument that we give was first given by O. Schreier in 1923. Consider the element $a^n = b^m$ in G where

$$G = \langle \{a,b\}, \{a^n = b^m\} \rangle.$$

This element commutes with a and b,

$$a\,a^n = a^{n+1} = a^n a, \quad ba^n = bb^m = b^{m+1} = b^m b = a^n b,$$

and so commutes with every element of G. The subgroup N generated by a^n is therefore a normal subgroup of G, and so the quotient group G/N is defined. If $g \in G$ then we may write g as

$$g = a^{\alpha(1)} b^{\beta(1)} \ldots a^{\alpha(k)} b^{\beta(k)}$$

for some $\alpha(1), \beta(1), \ldots, \alpha(k), \beta(k)$. Then $gN \in G/N$ may be written as

$$gN = (a^{\alpha(1)} N)(b^{\beta(1)} N) \ldots (a^{\alpha(k)} N)(b^{\beta(k)} N)$$
$$= (aN)^{\alpha(1)} (bN)^{\beta(1)} \ldots (aN)^{\alpha(k)} (bN)^{\beta(k)},$$

which shows that G/N is generated by aN and bN. If $gN = N$ then $g \in N$, and so $g = (a^n)^{\ell} = (b^m)^{\ell}$ for some ℓ. Thus the relations in G/N are given by $(aN)^n = N$ and $(bN)^m = N$, so that

$$G/N = \langle \{aN, bN\}; \{(aN)^n = 1, (bN)^m = 1\} \rangle$$
$$\cong \langle \{c, d\}; \{c^n = 1, d^m = 1\} \rangle.$$

Notice that G/N has a trivial centre because if x belongs to the centre of G/N then $cx = xc$ and $dx = xd$. The first condition implies that $x = c^{\alpha}$ for some α and the second implies that $x = d^{\beta}$ for some β; thus $x = 1$. This means that the centre $Z(G)$ of G is N (if $p: G \to G/N$ denotes the natural projection then it is easy to see that $p(Z(G)) \subseteq Z(G/N)$). In other words we have

$$G/Z(G) \cong \langle \{c, d\}; \{c^n = 1, d^m = 1\} \rangle.$$

Abelianizing this group gives

$$\langle \{c, d\}; \{c^n = 1, d^m = 1, cd = dc\} \rangle \cong \mathbb{Z}_n \times \mathbb{Z}_m.$$

Now if the torus knots of type (m,n) and (m',n') are similar then their groups G, G' are isomorphic. But then the groups $G/Z(G)$ and $G'/Z(G')$ are isomorphic. Abelianizing means that the groups $\mathbb{Z}_n \times \mathbb{Z}_m$ and $\mathbb{Z}_{n'} \times \mathbb{Z}_{m'}$ are isomorphic and this is possible only if $\{m, n\} = \{m', n'\}$ since the pairs m, n and m', n' are coprime.

27.7   Exercises

(a) Show that torus knots of type $(m, 1)$ or $(1, n)$ can be unknotted.

(b) Show that a torus knot of type (m, n) is similar to one of type (n, m). Are they equivalent?

(c) Torus knots were defined for positive pairs of coprime integers (m, n), but it makes sense to define torus knots for any pair of coprime (non-zero) integers. Show that if we change the sign of m and/or of n then the resulting torus knot is similar to the original one. Is it equivalent to the original one?

(d) Prove that if we abelianize the group of a torus knot then we get $\mathbb{Z}$. (Hint: If $G = \langle \{a, b\}; \{a^n = b^m\} \rangle$ then define $\varphi: AG \to \mathbb{Z}$ by $\varphi(a^k b^{\ell}) = mk + n\ell$.)

(e) Prove that the group $\langle \{a, b\}; \{a^3 = b^2\} \rangle$ is isomorphic to the group $\langle \{x, y\}; \{xyx = yxy\} \rangle$. (Use the Tietze transformations, Exercise 23.1(f).)

28

Knots: II Tame knots

Let K be a knot in $\mathbb{R}^3$. Let p denote the projection of $\mathbb{R}^3$ onto the plane $\mathbb{R}^2 = \{ (x_1,x_2,0) \in \mathbb{R}^3 \}$ given by $p(x_1,x_2,x_3) = (x_1,x_2,0)$. We say that $x \in p(K)$ is a *crossing point* if $p^{-1}(x) \cap K$ consists of more than one point. It is a double crossing point if $p^{-1}(x) \cap K$ consists of two points; see Figure 28.1. A double crossing point is *improper* if by moving the knot slightly the double crossing point disappears (in the illustration of an improper double crossing point in Figure 28.1 moving the 'top bit' to the left causes the double crossing point to disappear).

28.1 Definition

A knot K is *tame* if it is similar to a knot which has only finitely many crossing points each of which is a proper double crossing point.

Figure 28.1

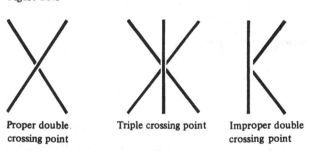

Proper double crossing point   Triple crossing point   Improper double crossing point

Figure 28.2. A wild knot.

All the examples of knots given in the last chapter are tame knots. An example of a knot which is not tame is given in Figure 28.2. Knots which are not tame are called *wild*. This chapter is concerned with tame knots.

28.2   Exercises

(a) Prove that a torus knot of type (m,n) is tame.

(b) Prove that a knot that has a finite number of crossing points is tame.

(c) Prove that a knot is tame if and only if it is similar to a knot that consists of a finite number of straight-line segments.

(d) Prove that a knot K is tame if and only if there exists a subspace $K_\epsilon \subset \mathbb{R}^3$ with $K \subset K_\epsilon$ such that K_ϵ is homeomorphic to $S^1 \times D^2$ with K corresponding to $S^1 \times \{0\}$ under this homeomorphism.

Throughout this chapter let K be a tame knot. Our object will be to calculate the group of K. We may assume that K lies in the lower half-space of $\mathbb{R}^3$, i.e.

$$K \subset \{(x_1,x_2,x_3); x_3 \leq 0\},$$

and furthermore that K lies in the plane $\{(x_1,x_2,x_3); x_3 = 0\}$ except where it dips down by a distance, say ϵ, at each (proper double) crossing point. See Figure 28.3.

Figure 28.3

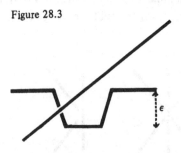

Let P denote the set of points in K of the form $(x_1,x_2,-\epsilon)$ for which $p(x_1,x_2,-\epsilon)$ is a crossing point. We can assume that $P \neq \emptyset$ since otherwise the knot K is unknotted. Let p_1 be one of the points in P. Putting an arrow (direction) on K determines $p_2,p_3,...,p_n$ where n is the number of crossing points. The set P divides K into a finite number of arcs $a_1,a_2,...,a_n$. The arrow on K gives one on each of $a_1,a_2,...,a_n$ and we may insist that the end point of a_i is p_i, $i = 1,2,...,n$. See Figure 28.4.

Our next object is to describe closed paths in $\mathbb{R}^3-K$. For $i = 1,2,...,n$ let c_i denote a small circle in $\mathbb{R}^3-K$ that goes around the arc a_i as indicated in

Figure 28.4

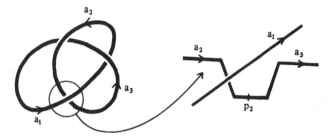

Figure 28.5

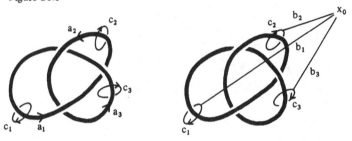

Figure 28.5. We give c_i an arrow so that together with the arrow on a_i we get a clockwise screw. Naturally we must choose the $c_1, c_2, ..., c_n$ to be disjoint.

Let x_0 be our base point in $\mathbb{R}^3 - K$, which is somewhere high above K. For $i = 1, 2, ..., n$ let b_i be a straight line in $\mathbb{R}^3 - K$ from x_0 to the circle c_i; the b_i should be chosen disjoint and each b_i should be in the upper half-space $\{ (x_1, x_2, x_3); x_3 \geq 0 \}$. See Figure 28.5. Define γ_i to be the closed path, based at x_0, that begins at x_0, goes along b_i, once along c_i in the direction indicated by the arrow on c_i and then back along b_i to x_0.

We shall show that $\pi(\mathbb{R}^3 - K, x_0)$ is generated by the elements $[\gamma_1]$, $[\gamma_2], ..., [\gamma_n]$. For the relations see Figure 28.6, which shows that if the crossing is as illustrated in Figure 28.6(a) then we have the relation

$$[\gamma_i] \, [\gamma_j] \, [\bar{\gamma}_{i+1}] \, [\bar{\gamma}_j] = 1.$$

The other possibility is that the arrow on a_j is opposite to that shown in Figure 28.6(a), namely as in Figure 28.6(c). In this case we have the relation

$$[\gamma_i] \, [\bar{\gamma}_j] \, [\bar{\gamma}_{i+1}] \, [\gamma_j] = 1.$$

It turns out that these are the only relations, as we shall show. In the above and throughout we let $a_{n+1} = a_1$ and $\gamma_{n+1} = \gamma_1$ etc.

Figure 28.6

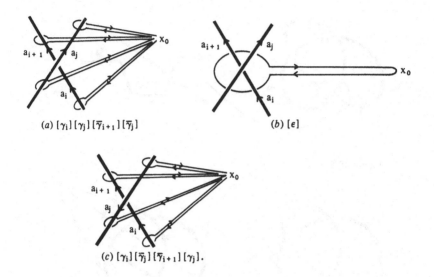

(a) $[\gamma_i][\gamma_j][\bar{\gamma}_{i+1}][\bar{\gamma}_j]$ (b) $[\epsilon]$

(c) $[\gamma_i][\bar{\gamma}_j][\bar{\gamma}_{i+1}][\gamma_j]$.

28.3 Theorem

The group of the tame knot K is generated by

$$[\gamma_1],[\gamma_2],...,[\gamma_n]$$

and has relations

$$[\gamma_1] = r_1, \ [\gamma_2] = r_2, \ ..., \ [\gamma_n] = r_n.$$

Each relation $[\gamma_i] = r_i$ is of the form

$$[\gamma_i] = [\gamma_j] \ [\gamma_{i+1}] \ [\gamma_j]^{-1}$$

or

$$[\gamma_i] = [\gamma_j]^{-1} \ [\gamma_{i+1}] \ [\gamma_j]$$

for some j. Moreover any one of the relations $[\gamma_k] = r_k$ may be omitted and the result remains true.

The j that appears in the relation is determined by the arc a_j that crosses over the point p_i. Which of the two relations holds depends on the arrows on a_i and a_j. In particular if the arrows determine locally a clockwise screw then the first relation holds, otherwise the second holds. See Figure 28.6(a) and (c) respectively.

By looking at the form of the relations given in Theorem 28.3 we immediately have:

28.4 Corollary

Abelianizing a knot group gives $\mathbb{Z}$.

To prove Theorem 28.3 we let C and A be subspaces of $\mathbb{R}^3$ defined by

$$C = \{ (x_1,x_2,x_3); x_3 > -2\epsilon/3 \},$$
$$A = \{ (x_1,x_2,x_3); x_3 < -\epsilon/3 \}.$$

The set $C \cap K$ (and $A \cap K$) consists of n disjoint arcs with no crossing points. It is therefore clear that there is a homeomorphism h: $C \to C$ such that $h(C \cap K)$ is the union

$$\bigcup_{i=1}^{n} (S_i \cap C)$$

where S_i, i = 1,2,...,n, is the circle $\{ (x_1,x_2,x_3); x_1 = i, x_2^2 + x_3^2 = 1 \}$. See Figure 28.7.

Figure 28.7

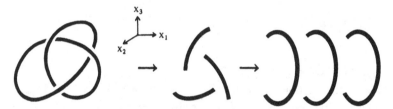

Thus C-K is homotopy equivalent to a disc with n points removed, from which it follows that $\pi(C-K, x_0)$ is a free group on n generators $[\gamma_1]$, $[\gamma_2]$,...,$[\gamma_n]$.

In a similar way we see that the fundamental group of A-K is a free group on n generators. However, A-K does not contain x_0, so let b be a straight line from x_0 to A, disjoint from K and all other curves chosen. Let

Figure 28.8

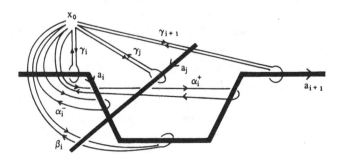

B be A together with all points of distance less than, say, δ from b. The space B-K is open and path connected and it is clear that $\pi(\text{B-K}, x_0)$ is a free group on n generators $[\beta_1], [\beta_2], ..., [\beta_n]$. The closed path β_i runs from x_0 along b to A, then to a point close to p_i, along a circle in B-K with centre p_i and then back again to x_0. See Figure 28.8.

Next we look at the space $(\text{B-K}) \cap (\text{C-K})$. This clearly has the homotopy type of a disc with 2n points removed and so $\pi((\text{B-K}) \cap (\text{C-K}), x_0)$ is a free group on 2n generators $[\alpha_1^-], [\alpha_2^-], ..., [\alpha_n^-], [\alpha_1^+], [\alpha_2^+], ..., [\alpha_n^+]$. To describe these generators note that $B \cap C$ splits K into 2n arcs which we may denote by $a_1^-, a_2^-, ..., a_n^-$ and $a_1^+, a_2^+, ..., a_n^+$ where a_i^-, a_i^+ are those parts of a_i, a_{i+1} respectively in $A \cap B$ that are nearest the point p_i. The path α_i^- goes around a_i^- whereas α_i^+ goes around a_i^- and a_i^+ (for convenience). See Figure 28.8.

Now let $\varphi_B: (B \cap C) - K \to B - K$ and $\varphi_C: (B \cap C) - K \to C - K$ denote the natural inclusions. The following equivalences are easy to see:

$$\varphi_B \, \alpha_i^- \sim \beta_i, \qquad \varphi_B \, \alpha_i^+ \sim \epsilon, \qquad \varphi_C \, \alpha_i^- \sim \gamma_i,$$

while

$$\varphi_C \, \alpha_i^+ \sim ((\gamma_i * \gamma_j) * \bar{\gamma}_{i+1}) * \bar{\gamma}_j$$

or

$$\varphi_C \, \alpha_i^+ \sim ((\gamma_i * \bar{\gamma}_j) * \bar{\gamma}_{i+1}) * \gamma_j,$$

depending on the relation of the arrow on a_j with the arrows on a_i and a_{i+1}. (For Figure 28.8 the relation is the first one given above.)

We may apply the Seifert-Van Kampen theorem since the spaces B-K, C-K, $(\text{B-K}) \cap (\text{C-K})$ are all open and path connected. The result concerning the generators and relations of $\pi(\mathbb{R}^3 - K, x_0)$ follows immediately.

To show that any one of the relations is redundant we simply alter A to A', where A' is A together with all points of distance greater than some large number, say N, from the origin of $\mathbb{R}^3$. This then gives us $B' = B \cup A'$. The fundamental group of B'-K is the same as B-K; however, the fundamental group of $(B' \cap C) - K$ has one fewer generator than that of $(B \cap C) - K$. This is because $(B' \cap C)$-K now has the homotopy type of a sphere S^2 with 2n points removed. So by removing the generator $[\alpha_k^+]$ from $\pi((B' \cap C) -K, x_0)$ we see that the relation $[\gamma_k] = r_k$ no longer is necessary in $\pi(\mathbb{R}^3 - K, x_0)$. We leave the details for the reader.

We illustrate the above theorem by three examples. For brevity we shall denote $[\gamma_k]$ simply by γ_k. First we recalculate the group of a trefoil knot. Using the notation in Figure 28.9 and previous notation we see that the group of a trefoil knot has three generators $\gamma_1, \gamma_2, \gamma_3$ with relations $\gamma_1 = \gamma_3^{-1} \gamma_2 \gamma_3$, $\gamma_2 = \gamma_1^{-1} \gamma_3 \gamma_1$, $\gamma_3 = \gamma_2^{-1} \gamma_1 \gamma_2$, any one of which is redundant.

Figure 28.9

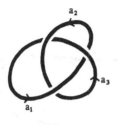

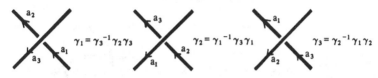

That γ_3, for example, is redundant can be seen simply by putting the first relation partly into the second to obtain the third, viz.

$$\gamma_2 = \gamma_1^{-1}\gamma_3\gamma_1 = \gamma_1^{-1}\gamma_3(\gamma_3^{-1}\gamma_2\gamma_3) = \gamma_1^{-1}\gamma_2\gamma_3,$$

so that $\gamma_3 = \gamma_2^{-1}\gamma_1\gamma_2$. Thus the group of a trefoil knot is

$$\langle\{\gamma_1,\gamma_2,\gamma_3\};\{\gamma_1 = \gamma_3^{-1}\gamma_2\gamma_3, \gamma_2 = \gamma_1^{-1}\gamma_3\gamma_1\}\rangle$$
$$=\langle\{\gamma_2,\gamma_3\};\{\gamma_2 = \gamma_3^{-1}\gamma_2^{-1}\gamma_3\gamma_3\gamma_3^{-1}\gamma_2\gamma_3\}\rangle$$
$$=\langle\{\gamma_2,\gamma_3\};\{\gamma_2\gamma_3\gamma_2 = \gamma_3\gamma_2\gamma_3\}\rangle$$

which can be seen (quite easily) to be isomorphic to the group

$$\langle\{a,b\};\{a^3 = b^2\}\rangle.$$

For the next example consider the granny knot of Figure 28.10(*a*). We see that the group of this knot has generators $\gamma_1,\gamma_2,...,\gamma_6$ and relations

$$\gamma_1 = \gamma_3^{-1}\gamma_2\gamma_3, \qquad \gamma_2 = \gamma_1^{-1}\gamma_3\gamma_1,$$
$$\gamma_3 = \gamma_2^{-1}\gamma_4\gamma_2, \qquad \gamma_4 = \gamma_6^{-1}\gamma_5\gamma_6,$$
$$\gamma_5 = \gamma_4^{-1}\gamma_6\gamma_4, \qquad \gamma_6 = \gamma_5^{-1}\gamma_1\gamma_5,$$

any one of which is redundant. All relations may be rewritten in terms of γ_1,γ_3 and γ_5 and it is not difficult to show that the above group is isomorphic to one on three generators $\gamma_1,\gamma_3,\gamma_5$ and two relations $\gamma_1\gamma_3\gamma_1 = \gamma_3\gamma_1\gamma_3$ and $\gamma_5\gamma_1\gamma_5 = \gamma_1\gamma_5\gamma_1$. Thus the group of the granny knot is the group on three generators x,y,z and two relations xyx = yxy and xzx = zxz.

For our last example consider the square knot of Figure 28.10(*b*). The group of this knot has six generators $\gamma_1,\gamma_2...,\gamma_6$ and relations

$$\gamma_1 = \gamma_3^{-1}\gamma_2\gamma_3, \qquad \gamma_2 = \gamma_1^{-1}\gamma_3\gamma_1,$$
$$\gamma_3 = \gamma_2^{-1}\gamma_4\gamma_2, \qquad \gamma_4 = \gamma_6\gamma_5\gamma_6^{-1},$$
$$\gamma_5 = \gamma_1\gamma_6\gamma_1^{-1}, \qquad \gamma_6 = \gamma_5\gamma_1\gamma_5^{-1},$$

Figure 28.10

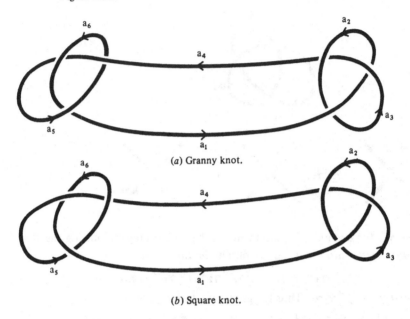

(a) Granny knot.

(b) Square knot.

any one of which is redundant. As in the granny knot it is easy to eliminate three of the generators to obtain the group with three generators $\gamma_1, \gamma_3, \gamma_5$ and two relations $\gamma_1 \gamma_3 \gamma_1 = \gamma_3 \gamma_1 \gamma_3$ and $\gamma_5 \gamma_1 \gamma_5 = \gamma_1 \gamma_5 \gamma_1$. Thus the group of a square knot is the group on three generators x,y,z, and two relations xyx = yxy and xzx = zxz. In particular we see that the groups of a granny knot and a square knot are isomorphic. It is however a fact that these two knots are not similar, although we shall not prove this.

28.5   Exercise

Using Theorem 28.3, calculate the groups of the knots in Figure 27.1. For the knot in Figure 27.1(b) show that the answer is $\mathbb{Z}$ as it should be.

We end our chapters on knots by briefly describing two constructions associated with knots. Most of the details and interesting properties of these constructions are left for the reader in the form of exercises.

Given two knots K_1, K_2 place an arrow on each and define their *connected sum* $K_1 \# K_2$ to be the knot obtained by removing an interval from each knot and glueing the result together so that the arrows go in the same direction; see Figure 28.11. A *prime knot* is one which cannot be expressed as $K_1 \# K_2$ where K_1 and K_2 are both not unknotted. Most tables of knots (see the appendix to Chapter 28) are of prime knots.

Figure 28.11

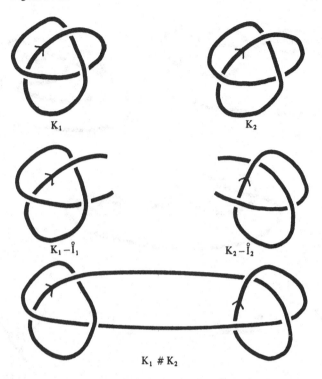

K_1

K_2

$K_1 - \mathring{I}_1$

$K_2 - \mathring{I}_2$

$K_1 \# K_2$

28.6 Exercises

(a) Show that $K_1 \# (K_2 \# K_3)$ is equivalent to $(K_1 \# K_2) \# K_3$.

(b) Show that $K_1 \# K_2$ is equivalent to $K_2 \# K_1$. (Hint: See Figure 28.12.)

(c) Show that if K_1 is equivalent to K_1' and K_2 is equivalent to K_2' then $K_1 \# K_2$ is equivalent to $K_1' \# K_2'$. Does the result still hold if the word equivalent is replaced by similar?

The next construction associates to each (tame) knot a surface-with-boundary (see Exercise 11.8(b)). In order to do this, first place an arrow on the knot. The region near each crossing point (Figure 28.13(a)) is replaced by the arrangement shown in Figure 28.13(b).

What remains is a number of disjoint circles. We may fill each circle with a disc in such a way that all the resulting discs are disjoint. To do this we may have to push the discs slightly off the plane in case the circles are nested; start with an innermost one and work outwards. Finally place a half-twisted

Figure 28.12

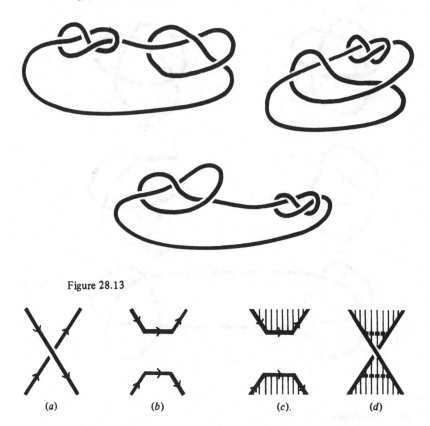

Figure 28.13

(a)   (b)   (c).   (d)

strip at the old crossing point (Figure 28.13(*d*)). The result is an orientable surface-with-boundary. Its boundary is of course the knot K. We say that the surface-with-boundary *spans* the knot. Some examples are given in Figures 28.14 and 28.15.

In Figure 28.15 a pair of nested circles arise. We have filled in the inner-most one by a disc in the plane. The outermost circle has been filled in with a disc that goes below the plane. Thus Figure 28.15(*b*) should be thought of as a sphere with a 'hole' in it and within this 'hole' there is a disc. For Figure 28.15(c) these two regions have been bridged by five half-twisted strips.

By the genus of a surface-with-boundary we mean the genus of the asso-ciated surface (in the sense of Exercise 11.8(b)). The genus of the surface-with-boundary that we have constructed is $\frac{1}{2}(c-d+1)$ where c is the number

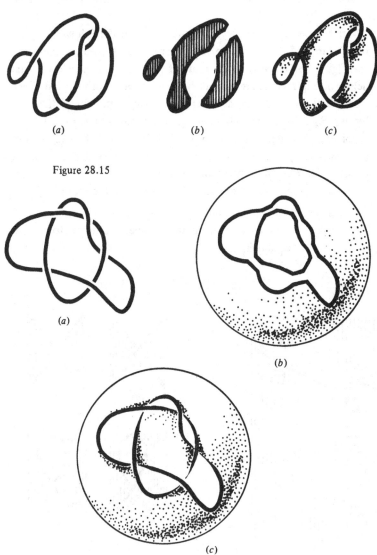

of crossing points and d is the number of disjoint circles that appeared in
our construction.

It is possible that there may be many different surfaces-with-boundary
spanning the knot. The *genus* of a knot K is defined to be the least integer
g(K) such that K is spanned by an orientable surface-with-boundary of genus
g(K).

Figure 28.14

(a)   (b)   (c)

Figure 28.15

(a)

(b)

(c)

28.7   Exercises

(a) Show that $g(K \# L) = g(K)+g(L)$. (This is not trivial.)

(b) Deduce from (a) that every knot can be written as a finite connected sum of prime knots.

(c) Prove that the genus of an unknotted knot is 0.

(d) Prove that if K is not unknotted then neither is $K \# L$ for any knot L.

(e) Prove that the genus of a torus knot of type (m,n) is less than or equal to $\frac{1}{2}(m-1)(n-1)$.

(f) Given a knot K 'drawn' in the plane, perform the following operations. First of all shade the largest area that surrounds the knot; see Figure 28.16. Next shade some of the regions in such a way that neighbouring regions are neither both white nor both shaded. Then label all the shaded regions except the largest one around the knot by $R_1, R_2,..., R_n$. At each crossing point put either $+1$, -1, or 0 according to whether the crossing is as in Figure 28.16(*b*), (*c*) or (*d*) respectively; the 0 occurs when the two shaded regions near the crossing point belong to the same region.

Form a symmetric $n \times n$ matrix $A(K) = (a_{ij})$ in the following way:

a_{ii} = sum of crossing point numbers at region R_i;

$-a_{ij} = -a_{ji}$ = sum of crossing point numbers common to regions R_i and R_j.

Figure 28.16

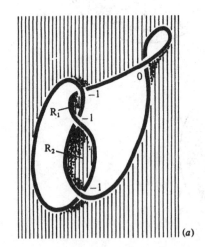

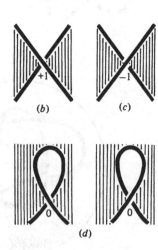

For example, the knot of Figure 28.16(*a*) determines the matrix

$$\begin{pmatrix} -2 & 1 \\ 1 & -2 \end{pmatrix}$$

Let $d(K) = \det A(K)$, so that $d(K)$ does not depend upon the labels on the shaded regions. If $n = 0$ define $d(K) = 1$.

(i) Find two equivalent knots K, L such that $d(K) \neq d(L)$. Is $|d(K)| = |d(L)|$?

(ii) If K and L are similar knots is $|d(K)| = |d(L)|$?

(iii) Find non-equivalent knots K, L such that $|d(K)| = |d(L)|$.

(iv) Show that $d(K \# L) = d(K)d(L)$.

The following diagrams show all prime knots, up to similarity, that have at most nine double crossing points.

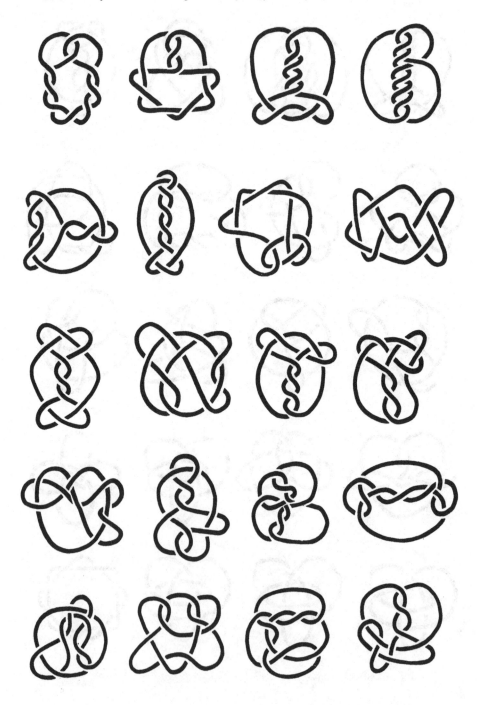

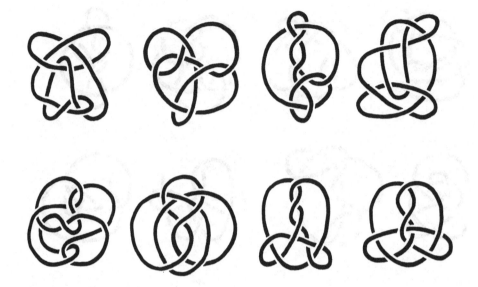

Singular homology: an introduction

Homology theory is without doubt very important in topology. This chapter cannot give the theory full justice. We shall merely illustrate the basic ideas involved and in a special case relate it to the fundamental group.

29.1   Definition

The *standard n-simplex* Δ_n is defined to be the following subspace of $\mathbb{R}^{n+1}$:

$$\Delta_n = \{\ x = (x_0, x_1, ..., x_n) \in \mathbb{R}^{n+1}\ ;\ \sum_{i=0}^{n} x_i = 1,\ x_i \geq 0,\ i = 0, 1, ..., n\ \}$$

The points $v_0 = (1, 0, ..., 0)$, $v_1 = (0, 1, 0, ..., 0)$,..., $v_n = (0, 0, ..., 0, 1)$ are called the *vertices* of Δ_n.

Figure 29.1

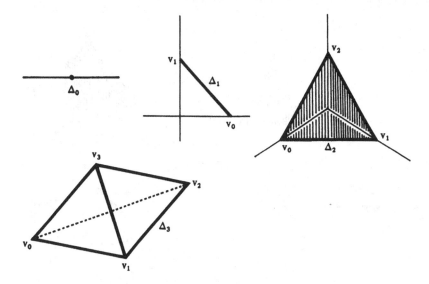

Thus Δ_0 is a single point, Δ_1 is a line interval, Δ_2 is a triangular region and Δ_3 is a solid tetrahedron; see Figure 29.1.

29.2   Definition

Let X be a topological space. A *singular n-simplex* in X is a continuous map $\varphi\colon \Delta_n \to X$.

Thus a singular 0-simplex is simply a point in X while a singular 1-simplex is essentially a path in X. Indeed if φ is a singular 1-simplex then defining $f(t) = \varphi(1-t, t)$ gives a path $f\colon I \to X$ from $\varphi(v_0)$ to $\varphi(v_1)$. Conversely, given a path $f\colon I \to X$ we obtain a singular 1-simplex $\varphi\colon \Delta_1 \to X$ by $\varphi(x_0, x_1) = f(x_1)$.

29.3   Definition

A *singular n-chain* in X is an expression of the form

$$\sum_{j \in J} n_j \varphi_j$$

where $\{\, \varphi_j;\, j \in J \,\}$ is the collection of all singular n-simplexes in X (with J some indexing set) and $n_j \in \mathbb{Z}$ with only a finite number of $\{\, n_j;\, j \in J \,\}$ being non-zero.

The set $S_n(X)$ of singular n-chains in X forms an abelian group with addition defined by

$$\sum n_j \varphi_j + \sum m_j \varphi_j = \sum (n_j + m_j) \varphi_j.$$

The zero element is $\sum 0\, \varphi_j$ and the inverse of $\sum n_j \varphi_j$ is $\sum (-n_j)\varphi_j$. Associativity is clear as is the fact that the resulting group is abelian.

Figure 29.2

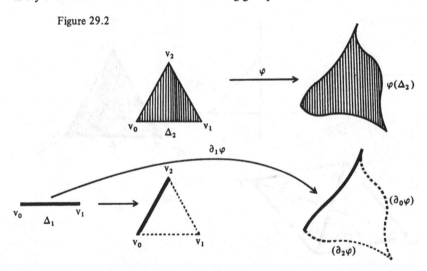

The group $S_n(X)$ has some nice properties but unfortunately it is in general extremely large. We make it more tractable by placing an equivalence relation on it (in a similar way that we used an equivalence relation to define the fundamental group). First we shall define the notion of a boundary operator.

Given a singular n-simplex φ define a singular (n-1)-simplex $\partial_i\varphi$ by

$$\partial_i\varphi(x_0,x_1,...,x_{n-1}) = \varphi(x_0,x_1,...,x_{i-1},0,x_i,...,x_{n-1})$$

for $i = 0,1,...,n$. See Figure 29.2. Clearly this leads to a homomorphism of groups

$$\partial_i: S_n(X) \to S_{n-1}(X),$$
$$\Sigma\, n_j\, \varphi_j \to \Sigma\, n_j\, \partial_i\, \varphi_j.$$

29.4   Definition

The *boundary operator* $\partial: S_n(X) \to S_{n-1}(X)$ is defined by

$$\partial = \partial_0 - \partial_1 + \partial_2 - ... + (-1)^n\, \partial_n = \sum_{i=0}^{n} (-1)^i \partial_i.$$

Using the boundary operator we can define two important subgroups of $S_n(X)$.

29.5   Definition

(a) A singular n-chain $c \in S_n(X)$ is an *n-cycle* if $\partial c = 0$. The set of n-cycles in X is denoted by $Z_n(X)$.

(b) A singular n-chain $d \in S_n(X)$ is an *n-boundary* if $d = \partial e$ for some $e \in S_{n+1}(X)$. The set of n-boundaries in X is denoted by $B_n(X)$.

In other words

$$Z_n(X) = \text{kernel } \partial: S_n(X) \to S_{n-1}(X),$$
$$B_n(X) = \text{image } \partial: S_{n+1}(X) \to S_n(X),$$

and so clearly both $Z_n(X)$ and $B_n(X)$ are subgroups of $S_n(X)$.

Notice that all singular 0-chains are 0-cycles, i.e. $Z_0(X) = S_0(X)$.

It turns out that all n-boundaries are n-cycles. This follows immediately from the next result.

29.6   Theorem

$$\partial\partial = 0.$$

Proof We check $\partial\partial$ on a singular n-simplex φ.

$$\partial\partial\varphi = \partial \sum_{i=0}^{n} (-1)^i \partial_i \varphi = \sum_{j=0}^{n-1} \sum_{i=0}^{n} (-1)^{i+j} \partial_j \partial_i \varphi.$$

Now if $i \leq j$ then we claim that $\partial_j \partial_i = \partial_i \partial_{j+1}$; we see this as follows.

$$
\begin{aligned}
(\partial_j \partial_i \varphi)(x_0,...,x_{n-2}) &= (\partial_j(\partial_i \varphi))(x_0,...,x_{n-2}) \\
&= (\partial_i \varphi)(x_0,...,x_{j-1},0,x_j,...,x_{n-2}) \\
&= \varphi(x_0,...,x_{i-1},0,x_i,...,x_{j-1},0,x_j,...,x_{n-2}) \\
&= (\partial_{j+1}\varphi)(x_0,...,x_{i-1},0,x_i,...,x_{n-2}) \\
&= (\partial_i \partial_{j+1}\varphi)(x_0,...,x_{n-2}).
\end{aligned}
$$

Thus

$$\partial\partial\varphi = \sum_{j=0}^{n-1} \sum_{i=0}^{j} (-1)^{i+j} \partial_j \partial_i \varphi + \sum_{j=0}^{n-1} \sum_{i=j+1}^{n} (-1)^{i+j} \partial_j \partial_i \varphi$$

$$= \sum_{j=0}^{n-1} \sum_{i=0}^{j} (-1)^{i+j} \partial_i \partial_{j+1} \varphi + \sum_{j=0}^{n-1} \sum_{i=j+1}^{n} (-1)^{i+j} \partial_j \partial_i \varphi$$

$$= \sum_{i=0}^{n-1} \sum_{j=i}^{n-1} (-1)^{i+j} \partial_i \partial_{j+1} \varphi + \sum_{j=0}^{n-1} \sum_{i=j+1}^{n} (-1)^{i+j} \partial_j \partial_i \varphi$$

$$= \sum_{j=0}^{n-1} \sum_{i=j}^{n-1} (-1)^{i+j} \partial_j \partial_{i+1} \varphi + \sum_{j=0}^{n-1} \sum_{i=j+1}^{n} (-1)^{i+j} \partial_j \partial_i \varphi$$

$$= \sum_{j=0}^{n-1} \sum_{i=j+1}^{n} (-1)^{i+j-1} \partial_j \partial_i \varphi + \sum_{j=0}^{n-1} \sum_{i=j+1}^{n} (-1)^{i+j} \partial_j \partial_i \varphi$$

$$= 0.$$

Thus $B_n(X)$ is a subgroup of $Z_n(X)$. Since both groups are abelian, $B_n(X)$ is a normal subgroup of $Z_n(X)$ and hence the quotient group $Z_n(X)/B_n(X)$ is defined.

29.7   Definition

The n-*th homology group* of X is defined as $Z_n(X)/B_n(X)$; it is denoted by $H_n(X)$.

In other words, elements of $H_n(X)$ are equivalence classes of cycles under the equivalence relation

$$c \sim c' \Leftrightarrow c - c' \in B_n(X)$$

for $c,c' \in Z_n(X)$. ($\sim$ is easily seen to be an equivalence relation.) In this case we say that c and c′ are *homologous* cycles.

Our next two results will determine the homology groups of a point and the zero homology group of a path connected space.

29.8   Lemma
    If X is a single point space then $H_0(X) \cong \mathbb{Z}$ and $H_n(X) = 0$ for $n > 0$.

Proof For all $n \geq 0$ there is a unique singular n-simplex $\varphi_{(n)} : \Delta_n \to X$ and so
$$S_n(X) = \mathbb{Z} = \{ k\varphi_{(n)}; k \in \mathbb{Z} \}.$$
Now $\partial_i \varphi_{(n)} = \varphi_{(n-1)}$ for $n > 0$ and

$$\partial \varphi_{(n)} = \sum_{i=0}^{n} (-1)^i \partial_i \varphi_{(n)} = \sum_{i=0}^{n} (-1)^i \varphi_{(n-1)}$$

$$= \begin{cases} 0 & \text{if n is odd,} \\ \\ \varphi_{(n-1)} & \text{if n is even and } n > 0. \end{cases}$$

For $n = 0$ we have $\partial \varphi_{(0)} = 0$.

From the above we see

$$Z_n(X) = \begin{cases} S_n(X) & \text{n odd or } n = 0, \\ 0 & \text{n even and } n > 0, \end{cases}$$

$$B_n(X) = \begin{cases} S_n(X) & \text{n odd,} \\ 0 & \text{n even,} \end{cases}$$

and hence

$$H_n(X) = \begin{cases} \mathbb{Z} & n = 0, \\ 0 & n > 0. \end{cases}$$

29.9   Lemma
    If X is a non-empty path connected space then $H_0(X) \cong \mathbb{Z}$.

Proof A typical 0-cycle (= a singular 0-chain) is of the form
$$\sum_{x \in X} n_x x$$

where $n_x \in \mathbb{Z}$ and only finitely many of $\{\, n_x; x \in X \,\}$ are non-zero. Define $\psi: H_0(X) \to \mathbb{Z}$ by

$$\psi(\Sigma\, n_x\, x) = \Sigma\, n_x.$$

First we check that this is well defined. Suppose that $\Sigma\, m_x\, x$ is another 0-cycle which is homologous to $\Sigma\, n_x\, x$, i.e.

$$\Sigma\, n_x\, x = \Sigma\, m_x\, x + \partial c$$

where c is a singular 1-chain. A singular 1-chain is of the form

$$c = \sum_{j \in J} k_j\, \varphi_j$$

where $k_j \in \mathbb{Z}$ and φ_j is a singular 1-simplex. Now

$$\partial c = \sum_{j \in J} k_j\, \partial\, \varphi_j = \sum_{j \in J} k_j(\varphi_j(v_1) - \varphi_j(v_0)),$$

so that

$$
\begin{aligned}
\psi(\Sigma n_x\, x) &= \psi(\Sigma m_x\, x + \partial c)\\
&= \psi(\Sigma m_x\, x + \Sigma k_j\, \varphi_j(v_1) - \Sigma k_j\, \varphi_j(v_0))\\
&= \Sigma m_x + \Sigma k_j - \Sigma k_j\\
&= \Sigma m_x\\
&= \psi(\Sigma m_x\, x),
\end{aligned}
$$

which shows that ψ is well defined.

Clearly ψ is a homomorphism. It is surjective because $\psi(nx) = n$, where x is any point of X. Finally we show that ψ is injective. Let $\Sigma\, n_x\, x$ be a 0-cycle; then

$$\Sigma\, n_x\, x = (\Sigma n_x)\, x_0 + \sum_{x \in X} (n_x\, x - n_x\, x_0)$$

$$= (\Sigma n_x)\, x_0 + \partial(\sum_{x \in X} n_x\, \varphi_x)$$

where φ_x is a path (= a singular 1-simplex) from x to x_0. Thus $\Sigma\, n_x\, x$ and $(\Sigma\, n_x)\, x_0$ are homologous. So if $\psi(\Sigma n_x\, x) = 0$ then $\Sigma n_x = 0$, and so Σn_x is homologous to 0, which proves that ψ is injective.

This last step is the crux of the matter because it shows that any 0-cycle $c = \Sigma n_x\, x$ is homologous to the 0-cycle $(\Sigma n_x)\, x_0$ which is completely determined by the integer Σn_x.

Given a continuous map f: $X \to Y$ then we may define

$$f_*: S_n(X) \to S_n(Y)$$

by

$$f_*(\sum_{j \in J} n_j\, \varphi_j) = \sum_{j \in J} n_j\, f\varphi_j$$

(Perhaps we should denote $f_{\#}$: $S_n(X) \to S_n(Y)$ by $f_{n\#}$: $S_n(X) \to S_n(Y)$ but this is unnecessarily complicated.) It is clear that $f_{\#}$ is a homomorphism of groups. In fact $f_{\#}$ sends cycles to cycles and boundaries to boundaries. This follows from the next result.

29.10 Lemma

$$\partial\, f_{\#} = f_{\#}\, \partial$$

Proof Consider a singular $(n-1)$-simplex φ. Then

$$
\begin{aligned}
((\partial_i f_{\#})(\varphi))(x_0,x_1,...,x_{n-1}) &= \partial_i(f\varphi)(x_0,x_1,...,x_{n-1}) \\
&= (f\varphi)(x_0,x_1,...,x_{i-1},0,x_i,...,x_{n-1}) \\
&= f(\varphi(x_0,x_1,...,x_{i-1},0,x_i,...,x_{n-1}) \\
&= f((\partial_i\varphi)(x_0,x_1,...,x_{n-1})) \\
&= (f\partial_i\varphi)(x_0,x_1,...,x_{n-1}) \\
&= ((f_{\#}\, \partial_i)(\varphi))(x_0,x_1,...,x_{n-1}),
\end{aligned}
$$

which proves the result.

29.11 Corollary

$$f_{\#}\, (Z_n(X)) \subseteq Z_n(Y),\ f_{\#}\, (B_n(X)) \subseteq B_n(Y).$$

Proof If c is a cycle in X then $\partial\, f_{\#}(c) = f_{\#}\, \partial(c) = 0$, which shows that $f_{\#}(c)$ is a cycle in Y. If d is a boundary in X then $d = \partial(e)$ and $f_{\#}(d) = f_{\#}\, \partial(e) = \partial f_{\#}(e)$ is a boundary in Y.

The above corollary shows that there is a homomorphism of groups

$$f_*: H_n(X) \to H_n(Y)$$

defined by

$$f_*\, (\sum_{j \in J}\, n_j\, \varphi_j) = \sum_{j \in J}\, n_j\, f\, \varphi_j$$

where $\sum_{j \in J}\, n_j\, \varphi_j$ is an n-cycle in X. We call $f_*: H_n(X) \to H_n(Y)$ the *induced homomorphism*.

The next two results are easy to prove and left for the reader. Compare Theorem 15.9 and Corollary 15.10.

29.12 Theorem

(i) Suppose $f: X \to Y$ and $g: Y \to Z$ are continuous maps; then $(gf)_* = g_* f_*: H_n(X) \to H_n(Z)$ for all $n \geq 0$.

(ii) If $1: X \to X$ is the identity map then 1_* is the identity homomorphism on $H_n(X)$ for all $n \geq 0$.

29.13   Corollary
    If f: $X \to Y$ is a homeomorphism then f_*: $H_n(X) \to H_n(Y)$ is an isomorphism for all $n \geq 0$.

Remark: Homology is a functor from topology to algebra (especially abelian groups). (See Remark preceding Exercises 15.11.)

    In fact if two spaces are homotopy equivalent then their homology groups are isomorphic. This follows from the next result, the *homotopy invariance theorem*.

29.14   Theorem
    Let f,g: $X \to Y$ be two continuous maps. If f and g are homotopic then $f_* = g_*$: $H_n(X) \to H_n(Y)$ for all $n \geq 0$.

Proof For $t \in I$ let λ_t: $X \to X \times I$ be given by $\lambda_t(x) = (x,t)$. Let F: $X \times I \to Y$ be a homotopy from f to g, i.e.

$$F(x,0) = f(x), \quad F(x,1) = g(x),$$

or in terms of λ_t we have

$$F \lambda_0 = f, \quad F \lambda_1 = g.$$

Suppose that $\lambda_{0*} = \lambda_{1*}$; then

$$f_* = (F\lambda_0)_* = F_* \lambda_{0*} = F_* \lambda_{1*} = (F\lambda_1)_* = g_*.$$

Thus we need only show that $\lambda_{0*} = \lambda_{1*}$: $H_n(X) \to H_n(X \times I)$. What we shall show is that for the homomorphisms

$$\lambda_{0\#}, \lambda_{1\#}: S_n(X) \to S_n(X \times I)$$

there exists a homomorphism (called the prism operator)

$$P: S_n(X) \to S_{n+1} (X \times I)$$

such that

$$\partial P + P \partial = \lambda_{1\#} - \lambda_{0\#}.$$

In such a situation the homomorphisms $\lambda_{0\#}$ and $\lambda_{1\#}$ are said to be *chain homotopic*.
    If $\lambda_{0\#}$ and $\lambda_{1\#}$ are chain homotopic and if c is an n-cycle in X then

$$(\lambda_{1\#} - \lambda_{0\#})(c) = (\partial P + P\partial)(c) = \partial (Pc)$$

which shows that $\lambda_{1\#}c$ and $\lambda_{0\#}c$ are homologous and hence $\lambda_{1*} = \lambda_{0*}$. Therefore in order to prove the theorem we have to show that $\lambda_{1\#}$ and $\lambda_{0\#}$ are chain homotopic. To do this we need to define P.
    Let φ: $\Delta_n \to X$ be a singular n-simplex in X, i.e. an element of $S_n(X)$. Let

$P_i(\varphi)$, for $i = 0, 1, ..., n$, be the element of S_{n+1} $(X \times I)$ defined by

$$P_i(\varphi)(x_0, x_1, ..., x_{n+1}) = \varphi(x_0, x_1, ..., x_{i-1}, x_i + x_{i+1}, x_{i+2}, ..., x_{n+1})$$
$$\times (1 - \sum_{k=0}^{i} x_k),$$

and let $P(\varphi) \in S_{n+1}$ $(X \times I)$ be given by

$$P(\varphi) = \sum_{i=0}^{n} (-1)^i P_i(\varphi).$$

It is not difficult to see that $P: S_n(X) \to S_{n+1}$ $(X \times I)$ is a homomorphism.

Now $\partial P(\varphi)$ may be rewritten as

$$\partial P(\varphi) = \sum_{j=0}^{n+1} (-1)^j \partial_j P(\varphi) = \sum_{j=0}^{n+1} \sum_{i=0}^{n} (-1)^{i+j} \partial_j P_i(\varphi).$$

We shall rewrite the $\partial_j P_i(\varphi)$ in another form.

If $i < j - 1$ then

$$\begin{aligned}
\partial_j P_i(\varphi)(x_0, ..., x_n) &= P_i(\varphi)(x_0, ..., x_{j-1}, 0, x_j, ..., x_n) \\
&= \varphi(x_0, ..., x_{i-1}, x_i + x_{i+1}, x_{i+2}, ..., x_{j-1}, 0, ..., x_n) \\
&\qquad \times (1 - \sum_{k=0}^{i} x_k) \\
&= \partial_{j-1} \varphi(x_0, ..., x_{i-1}, x_i + x_{i+1}, x_{i+2}, ..., x_n) \\
&\qquad \times (1 - \sum_{k=0}^{i} x_k) \\
&= P_i(\partial_{j-1} \varphi)(x_0, ..., x_n) \\
&= P_i \partial_{j-1}(\varphi)(x_0, ..., x_n).
\end{aligned}$$

If $i > j$ then

$$\begin{aligned}
\partial_j P_i(\varphi)(x_0, ..., x_n) &= P_i(\varphi)(x_0, ..., x_{j-1}, 0, x_j, ..., x_n) \\
&= \varphi(x_0, ..., x_{j-1}, 0, ..., x_{i-2}, x_{i-1} + x_i, x_{i+1}, ..., x_n) \\
&\qquad \times (1 - \sum_{k=0}^{i-1} x_k) \\
&= \partial_j \varphi(x_0, ..., x_{i-2}, x_{i-1} + x_i, x_{i+1}, ..., x_n) \\
&\qquad \times (1 - \sum_{k=0}^{i-1} x_k)
\end{aligned}$$

$$= P_{i-1} (\partial_j \varphi)(x_0,...,x_n)$$
$$= P_{i-1} \partial_j (\varphi) (x_0,...,x_n).$$

Finally if $i = j$ then

$$\partial_j P_j (\varphi) (x_0,...,x_n) = P_j (\varphi) (x_0,...,x_{j-1},0,x_j,...,x_n)$$

$$= \varphi(x_0,...,x_n) \times (1 - \sum_{k=0}^{j-1} x_k)$$

$$= P_{j-1} (\varphi) (x_0,...,x_{j-1},0,x_j,...,x_n)$$
$$= \partial_j P_{j-1} (\varphi) (x_0,...,x_n).$$

In summary we have

$$\partial_j P_j = \partial_j P_{j-1}$$
$$\partial_j P_i = P_{i-1} \partial_j \qquad \text{if } i > j$$
$$\partial_j P_i = P_i \partial_{j-1} \qquad \text{if } i < j-1.$$

By using these relations and by writing ∂P as

$$\partial P = \sum_{j=0}^{n+1} \sum_{i=0}^{n} (-1)^{i+j} \partial_j P_i$$

$$= \partial_0 P_0 + \sum_{i=j=1}^{n} \partial_j P_j + \sum_{i=j-1=0}^{n} (-1) \partial_j P_{j-1} - \partial_{n+1} P_n +$$

$$\sum_{i>j} (-1)^{i+j} \partial_j P_i + \sum_{i<j-1} (-1)^{i+j} \partial_j P_i$$

it is not difficult to see that

$$\partial P = \partial_0 P_0 - \partial_{n+1} P_n - P\partial.$$

But we have

$$\partial_0 P_0 (\varphi) (x_0,...,x_n) = P_0 (\varphi) (0,x_0,...,x_n)$$
$$= \varphi(x_0,...,x_n) \times 1$$
$$= \lambda_1 \varphi(x_0,...,x_n)$$
$$= \lambda_{1\#} (\varphi) (x_0,...,x_n),$$

and

$$\partial_{n+1} P_n(\varphi) (x_0,...,x_n) = P_n (\varphi) (x_0,...,x_n,0)$$
$$= \varphi(x_0,...,x_n) \times 0$$
$$= \lambda_0 \varphi(x_0,...,x_n)$$
$$= \lambda_{0\#} (\varphi) (x_0,...,x_n).$$

Thus

$$\partial P + P\partial = \lambda_{1\#} - \lambda_{0\#},$$

which shows that $\lambda_{1\#}$ and $\lambda_{0\#}$ are chain homotopic. This completes the proof of the theorem.

29.15 Exercises

(a) Prove that if f: $X \to Y$ is a homotopy equivalence then f_*: $H_n(X) \to H_n(Y)$ is an isomorphism for each $n \geq 0$.

(b) Prove that if i: $A \to X$ is the inclusion of a retract A of X then i_*: $H_n(A) \to H_n(X)$ is a monomorphism. Prove that if g: $X \to A$ is the retraction then

$$H_n(X) = \text{image } (i_*) \oplus \text{kernel } (g_*).$$

Prove furthermore that if A is a deformation retract of X then i_* is an isomorphism.

(c) Let X be a path connected space and let $x_0 \in X$ be some point of X. Let p: $X \to \{ x_0 \}$ denote the obvious map and define $\tilde{H}_n(X)$ to be the kernel of p_*: $H_n(X) \to H_n(\{ x_0 \})$. Prove that

$$H_n(X) \cong \tilde{H}_n(X) \oplus H_n(\{ x_0 \}).$$

(d) Let f: $X \to Y$ be a base point preserving continuous map. Show that there is an induced homomorphism

$$f_*: \tilde{H}_n(X) \to \tilde{H}_n(Y).$$

Suppose furthermore that g: $X \to Y$ is another base point preserving map with f and g homotopic relative to the base point of X. Prove that

$$f_* = g_*: \tilde{H}_n(X) \to \tilde{H}_n(Y).$$

The next result in our brief study of homology is a description of the relationship between the fundamental group and the first homology group of a space.

29.16 Theorem

There is a homomorphism

$$\psi: \pi(Y,y_0) \to H_1(Y).$$

If Y is path connected then ψ is surjective and the kernel of ψ is the commutator subgroup of $\pi(Y,y_0)$; in other words $H_1(Y)$ is $\pi(Y,y_0)$ abelianized.

Proof Suppose that f: $I \to Y$ is a path in Y that begins at y_0. Define $\psi(f)$: $\Delta_1 \to Y$ by

$$\psi(f) (x_0,x_1) = f(x_1) = f(1-x_0); (x_0,x_1) \in \Delta_1.$$

$\psi(f)$ is a singular 1-simplex. If f is a closed path then $\partial(\psi(f)) = y_0 - y_0 = 0$, and so $\psi(f)$ is a 1-cycle in Y.

We now check that if f and f' are equivalent closed paths then $\psi(f)$ and $\psi(f')$ are homologous cycles. Suppose that $f \sim f'$ and $F: I \times I \to Y$ is the homotopy relative to $\{0,1\}$ realizing the equivalence $f \sim f'$. We use F to define a singular 2-simplex $\varphi: \Delta_2 \to Y$ in the following way. The coordinates of a point Q in Δ_2 may be expressed as $(1-s, s(1-t), st)$ for some s,t with $0 \leq s, t \leq 1$ (see Figure 29.3). We define $\varphi(Q)$ to be $F(s,t)$. In terms of the coordinates (x_0, x_1, x_2) of $Q \in \Delta_2$ we have

$$\varphi(x_0, x_1, x_2) = \begin{cases} F(1-x_0, x_2/(1-x_0)) & \text{if } x_0 \neq 1, \\ F(0,0) & \text{if } x_0 = 1. \end{cases}$$

Notice that $x_2/(1-x_0) = x_2/(x_1+x_2)$ and of course $x_0, x_1, x_2 \geq 0$ so that

$$0 \leq x_2/(1-x_0) \leq 1.$$

Since $F(0,t) = F(0,0)$ for all $t \in I$ it follows that φ is continuous. The boundary of φ is easily calculated:

$$\partial_0 \varphi(x_0, x_1) = \varphi(0, x_0, x_1) = F(1, x_1) = y_0 = \psi(\epsilon)(x_0, x_1)$$

where $\epsilon: I \to Y$ is the constant path $\epsilon(t) = y_0$;

$$\partial_1 \varphi(x_0, x_1) = \varphi(x_0, 0, x_1) \quad = \begin{cases} F(1-x_0, x_1/(1-x_0)) & \text{if } x_0 \neq 1, \\ F(0,0) & \text{if } x_0 = 1 \end{cases}$$

$$= \begin{cases} F(1-x_0, 1) & \text{if } x_0 \neq 1, \\ F(0,0) & \text{if } x_0 = 1 \end{cases}$$

$$= \psi(f')(x_0, x_1);$$

$$\partial_2 \varphi(x_0, x_1) = \varphi(x_0, x_1, 0) \quad = F(1-x_0, 0) = \psi(f)(x_0, x_1).$$

Figure 29.3

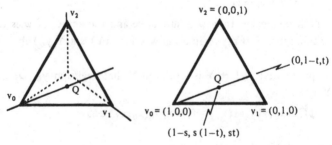

$v_2 = (0,0,1)$

$(0, 1-t, t)$

Q

$v_0 = (1,0,0)$ $v_1 = (0,1,0)$

$(1-s, s(1-t), st)$

In other words

$$\partial\varphi = \psi(f) + \psi(f') - \psi(\epsilon).$$

However, if we define $c_2 : \Delta_2 \to Y$ by $c_2(x_0, x_1, x_2) = y_0$ then we see that

$$\partial_0 c_2 = \partial_1 c_2 = \partial_2 c_2 = c_1$$

where $c_1 : \Delta_1 \to Y$ is given by $c_1(x_0, x_1) = y_0$. Thus $\psi(\epsilon) = \partial c_2$ and hence the cycles $\psi(f)$ and $\psi(f')$ are homologous. This proves that we have a well-defined function ψ from $\pi(Y, y_0)$ to $H_1(Y)$.

To check that ψ is a homomorphism let f, f' be two closed paths in Y, based at y_0. We need to show that $\psi(f * f')$ is homologous to $\psi(f) + \psi(f')$, i.e. that $\psi(f) + \psi(f') - \psi(f * f')$ is a boundary, say $\partial\varphi$ for some singular 2-simplex $\varphi : \Delta_2 \to Y$. The definition of φ is suggested by Figure 29.4 and is given explicitly by

$$\varphi(x_0, x_1, x_2) = \begin{cases} f(1 + x_2 - x_0) & \text{if } x_0 \geq x_2, \\ \\ f'(x_2 - x_0) & \text{if } x_0 \leq x_2. \end{cases}$$

Notice that φ is continuous by the glueing lemma. The boundary of φ can be easily calculated:

$$\partial_0 \varphi(x_0, x_1) = \varphi(0, x_0, x_1) = f'(x_1) = \psi(f')(x_0, x_1),$$

$$\partial_1 \varphi(x_0, x_1) = \varphi(x_0, 0, x_1) = \begin{cases} f(1 + x_1 - x_0) & \text{if } x_0 \geq x_1, \\ f'(x_1 - x_0) & \text{if } x_0 \leq x_1 \end{cases}$$

$$= \begin{cases} f(2x_1) & \text{if } x_1 \leq \frac{1}{2}, \\ f'(2x_1 - 1) & \text{if } x_1 \geq \frac{1}{2} \end{cases}$$

$$(x_1 + x_0 = 1)$$

$$= \psi(f * f')(x_0, x_1),$$

$$\partial_2 \varphi(x_0, x_1) = \varphi(x_0, x_1, 0) = f(1 - x_0) = \psi(f)(x_0, x_1).$$

Thus

$$\partial\varphi = \psi(f') - \psi(f * f') + \psi(f),$$

Figure 29.4

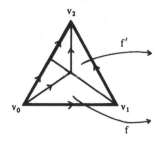

which shows that $\psi(f * f')$ is homologous to $\psi(f) + \psi(f')$ and hence that ψ is a homomorphism.

Suppose now that Y is path connected. We shall show that ψ is surjective. Let $c = \Sigma\, n_j\, \varphi_j$ be a 1-cycle in Y; thus $\partial c = 0$, i.e.

$$\Sigma\, n_j\, (\varphi_j(v_0) - \varphi_j(v_1)) = 0.$$

Rewriting ∂c as $\underset{y \in Y}{\Sigma}\, m_y\, y$, we must have $m_y = 0$ for all $y \in Y$. For each $j \in J$ choose a path g_{j_0} from y_0 to $\varphi_j(v_0) = \partial_0\, \varphi_j(1)$, and a path g_{j_1} from y_0 to $\varphi_j(v_1) = \partial_1\, \varphi_j(1)$. These paths must only depend upon the end point, thus if $\varphi_j(v_0) = \varphi_k(v_0)$ then $g_{j_0} = g_{k_0}$. Clearly we must have

$$\Sigma\, n_j\, (\psi(g_{j_1}) - \psi(g_{j_0})) = 0.$$

(Let g_y be a path from y_0 to y; then $\Sigma\, n_j(\psi(g_{j_1}) - \psi(g_{j_0}))$ may be rewritten as $\underset{y \in Y}{\Sigma}\, m_y\, \psi(g_y)$.)

Letting σ_j be the singular 1-chain defined by

$$\sigma_j = \psi(g_{j_0}) + \varphi_j - \psi(g_{j_1}),$$

then we have

$$c = \Sigma\, n_j\, \sigma_j.$$

If $f_j \colon I \to Y$ denotes the path given by $f_j(t) = \varphi_j(1-t,t)$ then $(g_{j_0} * f_j) * g_{j_1}$ is a closed path in Y, based at y_0, and

$$\psi((g_{j_0} * f_j) * g_{j_1}) = \sigma_j$$

and

$$\psi(\underset{j}{\Pi}\, [(g_{j_0} * f_j) * g_{j_1}]^{n_j}) = c,$$

which shows that ψ is surjective.

We prove that the kernel of ψ is the commutator subgroup. Suppose that $\psi(f)$ is homologous to 0, thus

$$\psi(f) = \partial(\underset{j \in J}{\Sigma}\, n_j\, \varphi_j) = \underset{j \in J}{\Sigma}\, n_j\, (\varphi_{j_0} - \varphi_{j_1} + \varphi_{j_2})$$

where φ_j $(j \in J)$ is a singular 2-simplex and $\varphi_{ji} = \partial_i \varphi_j$ $(i = 0,1,2)$. Since $\psi(f)$ is a singular 1-simplex we must have $\psi(f) = \varphi_{k\ell}$ for some k,ℓ and after collecting terms on the right hand side of the expression above $\psi(f) = \varphi_{k\ell}$ appears with coefficient 1 and all other terms have coefficient zero.

Let g_{ji} $(j \in J, i = 0,1,2)$ denote a path in Y from y_0 to $\varphi_j(v_i)$. As in the past g_{ji} is to depend only on the end point $\varphi_j(v_i)$ and not the particular indexing. If $\varphi_j(v_i) = y_0$ then choose the constant path. See Figure 29.5.

Let f_{ji} $(j \in J, i = 0,1,2)$ be paths in Y defined by

$$f_{ji}(t) = \varphi_{ji}(1-t,t) = \partial_i \varphi_i(1-t,t)$$

Figure 29.5

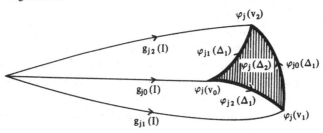

and define paths h_{ji} ($j \in J$, $i = 0,1,2$) by

$$h_{j_0} = (g_{j_1} * f_{j_0}) * \bar{g}_{j_2},$$
$$h_{j_1} = (g_{j_0} * f_{j_1}) * \bar{g}_{j_2},$$
$$h_{j_2} = (g_{j_0} * f_{j_2}) * \bar{g}_{j_1}.$$

Finally define paths h_j ($j \in J$) by

$$h_j = (h_{j_0} * \bar{h}_{j_1}) * h_{j_2}.$$

It is not difficult to see that h_j is equivalent to

$$(g_{j_1} * ((f_{j_0} * \bar{f}_{j_1}) * f_{j_2})) * \bar{g}_{j_1}$$

which is clearly equivalent to the constant path ϵ. Thus

$$\prod_j [h_j]^{n_j} = 1.$$

Let $A\pi(Y, y_0)$ denote the quotient of $\pi(Y, y_0)$ by the commutator subgroup (i.e. $A\pi(Y, y_0)$ is $\pi(Y, y_0)$ abelianized). If $[\alpha]$ is an element of $\pi(Y, y_0)$ we denote by $[[\alpha]]$ the corresponding element in $A\pi(Y, y_0)$. Since $\prod [h_j]^{n_j} = 1$ we have

$$\prod [[h_j]]^{n_j} = 1.$$

We know that $\psi(f) = \varphi_{k\ell}$ for some k, ℓ. It follows that $f = f_{k\ell}$ and (by our choice of g_{ji}) also that $f = h_{k\ell}$. Since $A\pi(Y, y_0)$ is abelian we may collect the terms in the expression $\prod [[h_j]]^{n_j}$ and deduce that

$$\prod [[h_j]]^{n_j} = [[f]].$$

Thus $[[f]] = 1$, i.e. $[f]$ belongs to the commutator subgroup. We see therefore that the kernel of ψ is contained in the commutator subgroup. On the other hand the fact that $H_1(Y)$ is abelian means that the kernel of ψ contains the commutator subgroup. This completes the proof of the theorem.

29.17 Exercises

(a) Show that $H_1(S^1) \cong \mathbb{Z}$ and that $H_1((S^1)^n) \cong \mathbb{Z}^n$.

(b) Give an example to show that if Y is not connected then $A\pi(Y,y_0)$ is not isomorphic to $H_1(Y)$.

(c) Calculate the first homology group of (i) an orientable surface of genus g, (ii) a non-orientable surface of genus g. Deduce that two surfaces S_1,S_2 are homeomorphic if and only if $H_1(S_1) \cong H_1(S_2)$.

(d) Suppose that Y is path connected. Prove that $\pi(Y,y_0)$ and $H_1(Y)$ are isomorphic if and only if $\pi(Y,y_0)$ is abelian.

(e) Show that the first homology group of a figure 8 is isomorphic to $\mathbb{Z} \times \mathbb{Z}$.

(f) Let S be a surface and let S' be S with an open disc neighbourhood removed. Prove that $H_1(S) \cong H_1(S')$.

When calculating fundamental groups we found the Seifert–Van Kampen theorem very profitable. For homology theory we have an analogous theorem which we shall describe.

Let $X = U_1 \cup U_2$ where U_1 and U_2 are open subsets of X, and let φ_i: $U_1 \cap U_2 \to U_i$, ψ_i: $U_i \to X$ denote the inclusion maps for i = 1,2. Define homomorphisms

$$i : H_k(U_1 \cap U_2) \to H_k(U_1) \oplus H_k(U_2)$$
$$j : H_k(U_1) \oplus H_k(U_2) \to H_k(X)$$

by

$$i(c) = (\varphi_{1*}(c), \varphi_{2*}(c)),$$
$$j(c_1,c_2) = \psi_{1*}(c_1) - \psi_{2*}(c_2).$$

29.18   Theorem

Let $X = U_1 \cup U_2$ where U_1 and U_2 are open subsets of X. There are homomorphisms

$$\Delta : H_k(X) \to H_{k-1}(U_1 \cap U_2)$$

such that in the following sequence of groups and homomorphisms,

$$... \to H_{k+1}(X) \xrightarrow{\Delta} H_k(U_1 \cap U_2) \xrightarrow{i} H_k(U_1) \oplus H_k(U_2) \xrightarrow{j} H_k(X) \xrightarrow{\Delta} H_{k-1}(U_1 \cap U_2) \to ...$$

the kernel of each homomorphism is equal to the image of the preceding one.

Furthermore if Y is another space with $Y = V_1 \cup V_2$ (V_1,V_2 open in Y) and if f: $X \to Y$ is a continuous map with $f(U_i) \subseteq V_i$ then

$$(f|U_1 \cap U_2)_* \Delta = \Delta f_*,$$

i.e. the homomorphisms Δ commute with induced homomorphisms.

The homomorphisms Δ are called *connecting homomorphisms* and the sequence in Theorem 29.18 is called the *Mayer–Vietoris sequence*. In general a sequence of groups and homomorphisms in which the kernel of each homomorphism is equal to the image of the preceding one is called an *exact sequence*. Thus the Mayer–Vietoris sequence is an exact sequence.

We shall not prove Theorem 29.18 although we will indicate its usefulness (and hence the usefulness of homology theory) by proving a result and then deducing several important corollaries.

29.19 Theorem

Let n be a positive integer; then

$$H_k(S^n) = \begin{cases} \mathbb{Z} & \text{if } k = 0, n, \\ 0 & \text{otherwise.} \end{cases}$$

Moreover, if $T_n: S^n \to S^n$ is the reflection map given by $T_n(x_0, x_1, ..., x_n) = (-x_0, x_1, ..., x_n)$ then

$$T_{n*}: H_n(S^n) \to H_n(S^n)$$

is multiplication by -1.

Proof We prove the result inductively using the Mayer–Vietoris sequence. Let $U_1 = \{ x \in S^n; x_n > -\frac{1}{2} \}$ and $U_2 = \{ x \in S^n; x_n < \frac{1}{2} \}$. Note that U_1 and U_2 are contractible and that $U_1 \cap U_2$ is homotopy equivalent to S^{n-1} so that

$$H_k(U_i) = \begin{cases} \mathbb{Z} & \text{if } k = 0, \\ 0 & \text{otherwise;} \end{cases}$$

$$H_k(U_1 \cap U_2) = H_k(S^{n-1}).$$

Note that if we think of S^{n-1} as $\{ x \in S^n; x_n = 0 \}$ then $T_n|S^{n-1} = T_{n-1}$. Let $n=1$; then for $k=1$ the Mayer–Vietoris sequence becomes

$$... \to 0 \xrightarrow{j} H_1(S^1) \xrightarrow{\Delta} H_0(S^0) \xrightarrow{i} \mathbb{Z} \oplus \mathbb{Z} \xrightarrow{j} ...$$

which becomes

$$... \to 0 \xrightarrow{j} H_1(S^1) \xrightarrow{\Delta} \mathbb{Z} \oplus \mathbb{Z} \xrightarrow{i} \mathbb{Z} \oplus \mathbb{Z} \xrightarrow{j} ...$$

with $i(x,y) = (x+y, x+y)$. Now Δ is injective since $\ker(\Delta) = \text{im}(j) = 0$; furthermore $\text{im}(\Delta) = \ker(i) = \{ (x,-x) \in \mathbb{Z} \oplus \mathbb{Z} \}$ which is isomorphic to $\mathbb{Z}$ so that $H_1(S^1) = \mathbb{Z}$. It is clear that $T_{0*}(x,y) = (y,x)$ and since $T_{0*}\Delta = \Delta T_{1*}$ we see that T_{1*} is multiplication by -1. For $k > 1$ the sequence is

$$... \to 0 \xrightarrow{i} H_k(S^1) \xrightarrow{\Delta} H_{k-1}(S^0) \xrightarrow{j} 0 \to ...$$

and it is easily seen that Δ is an isomorphism (it is injective because $\ker(\Delta) = \operatorname{im}(i)$ and surjective because $\operatorname{im}(\Delta) = \ker(j)$). The theorem is therefore proved for $n=1$.

Suppose that $m > 1$ and that the result in question is true for $n=m-1$; then we shall show that it is also true for $n=m$.

If $k=1$ then we have

$$\ldots \to 0 \xrightarrow{j} H_1(S^m) \xrightarrow{\Delta} H_0(S^{m-1}) \xrightarrow{i} \mathbb{Z} \oplus \mathbb{Z} \to \ldots$$

which becomes

$$\ldots \to 0 \xrightarrow{j} H_1(S^m) \xrightarrow{\Delta} \mathbb{Z} \xrightarrow{i} \mathbb{Z} \oplus \mathbb{Z} \to \ldots$$

with $i(a) = (a,a)$ so that $\ker(i) = 0$ and hence $\operatorname{im}(\Delta) = 0$ and $H_1(S^m) = 0$.

If $k > 1$ then we have

$$\ldots \to 0 \xrightarrow{j} H_k(S^m) \xrightarrow{\Delta} H_{k-1}(S^{m-1}) \xrightarrow{i} 0$$

from which we deduce that $H_k(S^m) \cong H_{k-1}(S^{m-1})$. Furthermore if $k=m$ then using the fact that $T_{m-1 *}\Delta = \Delta T_{m *}$ we deduce that $T_{m *}$ is multiplication by -1. The result follows by induction.

29.20 Corollary

(a) If $n \neq m$ then S^n and S^m do not have the same homotopy type.

(b) Any continuous map $f: D^n \to D^n$ has a fixed point.

(c) The reflection map $T_n: S^n \to S^n$ is not homotopic to the identity map.

(d) The antipodal map $A: S^{2n} \to S^{2n}$ given by $A(x) = -x$ is not homotopic to the identity map.

(e) If $f: S^{2n} \to S^{2n}$ is homotopic to the identity then f has a fixed point.

(f) There is no continuous map $f: S^{2n} \to S^{2n}$ such that the vectors x and $f(x)$ are orthogonal in $\mathbb{R}^{2n+1}$ for all x.

Part (a) follows from the homotopy invariance theorem (Theorem 29.14); see Exercise 29.15(a). Part (b) is *Brouwer's fixed point theorem* and is proved in the same way as Corollary 16.10. Part (c) follows from the homotopy invariance theorem. Part (d) follows from the fact that $A = R_0 R_1 \ldots R_{2n}$ where R_i is reflection in the i-th coordinate so that $A_*: H_{2n}(S^{2n}) \to H_{2n}(S^{2n})$ is multiplication by $(-1)^{2n+1} = -1$. For part (e) assume that f has no fixed point so that $(1-t) f(x) - tx \neq 0$ for all x and so we may define a homotopy $F: S^{2n} \times I \to S^{2n}$ between f and A by

$$F(x,t) = ((1-t) f(x) - tx)/\|(1-t) f(x) - tx\|.$$

Finally part (f) follows from (e) because if x and f(x) are orthogonal then f(x) ≠ x.

Parts (e) and (f) have a physical interpretation for n = 1 which is commonly called the *hairy-ball theorem*. This states that if you have a hairy-ball (i.e. D^3 with a hair growing out from each point of the surface S^2) then you cannot comb it smoothly; indeed, any such attempt produces bald spots or partings in the hair. For the proof just observe that if you had a smoothly combed hairy-ball then the direction vector f(x) of the hair at x is orthogonal to the vector x. Note however that it is possible to comb a hairy-torus smoothly; this has important implications in nuclear fusion power stations.

29.21 Exercises

(a) Use the Mayer–Vietoris sequence to calculate the homology of $\mathbb{R} P^2$.

(b) Use the Mayer–Vietoris sequence to calculate the homology groups of the complement of a knot. Deduce Corollary 28.4.

(c) Prove that there is no retraction of D^n on S^{n-1}.

(d) Let M be an m-manifold and N an n-manifold. Prove that if m ≠ n then M and N are not homeomorphic. This is called the *topological invariance of dimension*. (Hint: Use the homeomorphism $M/(M-D) \cong S^m$ described in Exercise 11.12(f).)

There are many other different ways of defining homology groups. For a very large class of spaces (the CW complexes for example) all these theories coincide. This leads to an axiomatic approach to homology theory which was originated by S. Eilenberg and N. Steenrod in the early 1950s. We shall describe a set of axioms for so-called 'reduced homology theories'. These are theories defined on topological spaces with a base point (as is the fundamental group). The 'reduced singular homology groups' of a space X with base point $x_0 \in X$ are defined by

$$\tilde{H}_n(X) = \text{kernel}\,(p_*\colon H_n(X) \to H_n(\{\,x_0\,\}))$$

where p: $X \to \{\,x_0\,\}$ is the obvious map (see Exercise 29.15(c)). Given a base point preserving map f: $X \to Y$ there is an induced homomorphism f_*: $\tilde{H}_n(X) \to \tilde{H}_n(Y)$ defined in the obvious way.

Before giving the axioms of a reduced homology theory we briefly introduce some notation.

29.22 Definition

Let X be a topological space with base point x_0. Define ΣX to be the quotient space

$$(X \times I)/(X \times \partial I \cup \{\,x_0\,\} \times I)$$

with the obvious base point. We call ΣX the *(reduced) suspension* of X.

Notice that if f: X → Y is a base point preserving continuous map then it induces a base point preserving continuous map Σf: ΣX → ΣY defined in the obvious way.

29.23   Definitions

The *reduced cone* CX of X is defined to be the quotient space

$$(X \times I)/(X \times \{1\} \cup \{x_0\} \times I).$$

If f: X → Y is a base point preserving continuous map then the *mapping cone* C_f of f is the quotient space

$$(CX \cup Y)/\sim$$

where $\sim$ is the equivalence relation given by

$$(x,0) \sim f(x)$$

for $(x,0) \in CX$ and $f(x) \in Y$. The base point of C_f is the point in C_f corresponding to the base point y_0 of Y.

Note that there is a natural inclusion

$$i: Y \to C_f.$$

29.24   Exercises

(a) Prove that if p: X → $\{x_0\}$ is the constant map then C_p is just ΣX.

(b) Prove that if X is Hausdorff then so is ΣX.

(c) Prove that $\Sigma S^1 \cong S^2$.

We now give the Eilenberg–Steenrod axioms for a reduced homology theory. From now on all spaces have a base point and all maps between such spaces are continuous base point preserving maps.

A *reduced homology theory* defined on a collection of (possibly all) topological spaces with a base point consists of the following.

(A) A family $\{\tilde{H}_n; n \in Z\}$ such that $\tilde{H}_n$ assigns to each space X under consideration an abelian group $\tilde{H}_n(X)$. This group is called the *n-th reduced homology group* of X.

(B) For every base point preserving continuous map f: X → Y there is an *induced homomorphism* $f_*: \tilde{H}_n(X) \to \tilde{H}_n(Y)$ for all n.

(C) For each space X and each integer n there is a homomorphism $\sigma_n(X)$: $\tilde{H}_n(X) \to \tilde{H}_{n+1}(\Sigma X)$. The homomorphism $\sigma_n(X)$ is called the *suspension homomorphism*.

The above are subject to the following seven axioms.

(1) (The identity axiom) If $1: X \to X$ is the identity map then the induced homomorphism

$$1_*: \tilde{H}_n(X) \to \tilde{H}_n(X)$$

is an isomorphism for each integer n.

(2) (The composition axiom) If $f: X \to Y$ and $g: Y \to Z$ are (base point preserving continuous) maps then $(gf)_* = g_* f_*$.

(3) (The naturality of suspension axiom) If $f: X \to Y$ is a continuous map then the following diagram is commutative.

$$
\begin{array}{ccc}
\tilde{H}_n(X) & \xrightarrow{\ \sigma_n(X)\ } & \tilde{H}_{n+1}(\Sigma X) \\
\Big\downarrow{\scriptstyle f_*} & & \Big\downarrow{\scriptstyle (\Sigma f)_*} \\
\tilde{H}_n(Y) & \xrightarrow{\ \sigma_n(Y)\ } & \tilde{H}_{n+1}(\Sigma Y)
\end{array}
$$

(4) (The homotopy axiom) If the maps $f, g: X \to Y$ are homotopic relative to the base point of X then the induced homomorphisms f_* and g_* are equal.

(5) (The suspension axiom) The suspension homomorphism $\sigma_n(X): \tilde{H}_n(X) \to \tilde{H}_{n+1}(\Sigma X)$ is an isomorphism for all X and all n.

(6) (The exactness axiom) For every map $f: X \to Y$ the sequence

$$\tilde{H}_n(X) \xrightarrow{\ f_*\ } \tilde{H}_n(Y) \xrightarrow{\ i_*\ } \tilde{H}_n(C_f)$$

has the property that image (f_*) = kernel (i_*) for all n, where $i: Y \to C_f$ is the natural inclusion.

(7) (The dimension axiom)

$$
\tilde{H}_n(S^0) = \begin{cases} \mathbb{Z} & \text{if } n = 0, \\ 0 & \text{otherwise.} \end{cases}
$$

29.25 Exercise

Show that reduced singular homology theory is a reduced homology theory in the above sense. (Hint: For (5) and (6) use the Mayer–Vietoris sequence.)

If instead of the dimension axiom above we have

$$\tilde{H}_n(S^0) = G_n$$

for some collection of abelian groups G_n, $n \in \mathbb{Z}$ then we have what is called a *generalized reduced homology theory with coefficients* $\{ G_n; n \in \mathbb{Z} \}$. Such theories have earned a prominent position in modern algebraic topology.

30

Suggestions for further reading

This chapter contains a selection of books suitable for further reading. Books that assume far more knowledge of topology than is contained in this book have not been included. The choice of books given is based upon the author's (biased) preferences. The book [Spanier] is mentioned on a number of occasions; it is an excellent all-round book on algebraic topology, although some find it hard to read.

Manifolds For some general theory about manifolds see [Dold]. For 2-manifolds and 3-manifolds see [Moise]. An important class of manifolds is the so-called 'differentiable manifolds'; for a book about these see [Hirsch].

Homotopy theory Three recommended books are [Gray], [Spanier] and [Whitehead].

Covering spaces Covering spaces lead to 'fibre bundles' and good books for this are [Husemoller] and [Spanier].

Group actions On topological spaces see [Bredon]. On manifolds see [Conner & Floyd] and [Conner].

Knot theory [Rolfsen].

Homology theory For further singular homology theory see [Dold], [Greenberg], [Spanier] and [Vick]. Two other types of homology theory are simplicial homology and Cech homology. [Spanier] deals with both. [Maunder] is good for simplicial homology while [Dold] and [Massey] can be recommended for Cech homology theory. For generalized homology theories see [Gray] and [Switzer]. The book [Gray] deals with generalized homology theory from a purely homotopy theoretic point of view. Finally, the original book on axiomatic homology theory is [Eilenberg & Steenrod].

The books suggested

Bredon, G.E. *Introduction to compact transformation groups.* Academic Press, New York – London, 1972.

Conner, P.E. *Differentiable periodic maps* (second edition). Springer, Berlin – Heidelberg – New York, 1979.

Conner, P.E. & Floyd, E.E. *Differentiable periodic maps.* Springer, Berlin – Heidelberg – New York, 1964.

Dold, A. *Lectures on algebraic topology.* Springer, Berlin – Heidelberg – New York, 1972.

Eilenberg, S. & Steenrod, N. *Foundations of algebraic topology.* Princeton University Press, Princeton, N.J., 1952.

Gray, B. *Homotopy theory.* Academic Press, New York – San Francisco – London, 1975.

Greenberg, M.J. *Lectures on algebraic topology.* Benjamin, New York, 1967.

Hirsch, M.W. *Differential topology.* Springer, New York – Heidelberg – Berlin, 1976.

Husemoller, D. *Fibre bundles* (second edition). Springer, New York – Heidelberg – Berlin, 1975.

Massey, W.S. *Homology and cohomology theory.* Marcel Dekker, New York – Basel, 1978.

Maunder, C.R.F. *Introduction to algebraic topology.* Cambridge University Press, 1980.

Moise, E.E. *Geometric topology in dimensions 2 and 3.* Springer, New York – Heidelberg – Berlin, 1977.

Rolfsen, D. *Knots and links.* Publish or perish, Berkeley, Ca., 1976.

Spanier, E.H. *Algebraic topology.* McGraw Hill, New York, 1966.

Switzer, R.M. *Algebraic topology – homotopy and homology.* Springer, Berlin – Heidelberg – New York, 1975.

Vick, J.W. *Homology theory.* Academic Press, New York – London, 1973.

Whitehead, G.W. *Homotopy theory.* M.I.T. Press, Cambridge, Mass., 1966.

INDEX

abelian group, 4
abelianization of a group, 206
abelianized knot group, 225
action of a group, 35
    free action, 71
action of the fundamental group, 152
antipodal map, 256
antipodal points, 64
arc, 77, 92
    open, 20
arcwise connected space, 93
associativity, 3
axioms of homology theory, 258

ball, 13
base point, 124
base space of a covering, 143
bijective function, 2
binary operation, 3
Borsuk–Ulam theorem, 157
boundary
    of a chain, 241
    of a manifold-with-boundary, 91
    of a space, 15
boundary operator, 241
bounded subset of $\mathbb{R}^n$, 48
bowline knot, 210
Brouwer's fixed point theorem, 140, 256

cartesian product, 1
centre of a group, 126
chain, 240
chain homotopic maps, 246
Chinese button knot, 210
chord of a Jordan curve, 107
circle, 21
    Polish, 165
class, 2
classification theorem of surfaces, 80, 90
closed map, 17
closed path, 124
closed set, 13

closure, 14
comb (flea and comb), 95
commutative group, 4
commutator subgroup, 4
compactification, 49
compact-open topology, 49
compact space, 45
    locally compact, 49
    one-point compactification, 49
component, 100
composite function, 2
composition axiom, 259
concrete topology, 11
cone
    reduced, 258
    mapping, 258
connected space, 58
    locally path connected, 98, 162
    path connected, 93
    semilocally simply connected, 171
    simply connected, 130
connected sum
    of knots, 228
    of n-manifolds, 88
    of surfaces, 79
connecting homomorphism, 254
consequence (of relation), 178
constant map, 16
constant path, 92
continuous function
    between euclidean spaces, 6
    between metric spaces, 8
    between topological spaces, 16
    uniformly continuous, 103
continuous on the right, 17
contractible space, 114
convex subset of $\mathbb{R}^n$, 93
coset, 3
cover, 44
    finite, 44
    open, 45
    subcover, 44